3

Approximation Theory
and
Numerical Methods

Approximation Theory
and
Numerical Methods

G. A. Watson

Department of Mathematics
University of Dundee

A Wiley–Interscience Publication

JOHN WILEY & SONS
Chichester · New York · Brisbane · Toronto

British Library Cataloguing in Publication Data:

Watson, G Alistair
  Approximation theory and numerical methods.
  1. Approximation theory
  2. Numerical analysis
  I. Title
  511'.42   QA297.5   79–42725

  ISBN 0 471 27706 1

Photo typeset in India by The Macmillan Co. of India Ltd., Bangalore – 1
Printed in the United States of America by Vail-Ballou Press, Inc.,
Binghamton, N.Y.

*To
Hilary*

Contents

Preface

In the last 15 or so years, there has been considerable activity in the development of methods for solving problems in approximation theory. This effort has, in part, been a consolidation of traditional techniques, designed in many cases for important, though sometimes fairly restricted, subclasses of approximation problems. The availability of large-scale computing facilities, however, has also helped to encourage the construction of methods which are to a great extent free from the various classical restrictions. The study of these techniques is now part of many advanced numerical analysis courses, and the purpose of this book is to provide a text which combines the theory with a modern treatment of computational methods for solving standard approximation problems.

A fundamental choice facing the writer of a book on approximation is whether to treat the theory in classical style, with the methods and language of the theory of functions, or whether to use the modern methods of functional analysis. While adopting the first option in order to ensure that the material is accessible to the widest possible audience, I have attempted here to make some provision for those capable of appreciating the latter approach: included is a general treatment of the characterization of best approximations in normed linear spaces, and the derivation from this of specific theorems. This development may be omitted by readers without the necessary background, who can arrive at the relevant results via the more traditional routes.

Much of this book is based on courses of lectures given to Senior Honours students and to M.Sc. students at the University of Dundee. The general level of the text is that of a second course in numerical analysis, and a prerequisite is the material usually covered in a first (introductory) course. A familiarity is assumed with the basic material to be found in good courses in calculus, real analysis, and linear algebra, so that, for example, such topics as sequences, convergence, metric spaces, dimension, basis, vectors, and matrices are understood. It is also necessary in Chapter 7 to introduce the idea of a measure, although an intuitive understanding of the concept will be sufficient. Only real-valued functions of real variables are treated, so a knowledge of complex variable theory is not required. However, a familiarity with linear programming theory is necessary to understand fully some of the methods presented; although perhaps not an entirely reasonable assumption, I felt that the inclusion of a full treatment of the necessary linear programming results could not be justified. The required material can be obtained in, for example, the book by Hadley (1962).

After an introductory chapter, the next six chapters are devoted to the treatment of standard linear problems with the different L_p norms. Chapter 8 is concerned with piecewise polynomials, and is an introduction to what is a very important and still rapidly growing subject. The last two chapters are devoted to nonlinear problems, with Chapter 9 dealing in particular with the case of rational approximation. While the nature of this book is such that a very general treatment of the material is given throughout, I have tried not to ignore entirely those students whose aims are rather more modest. In particular, in Chebyshev approximation the treatment of the polynomial (or Chebyshev set) case is sometimes sufficient, and it is not always convenient to extract this as a special case of a more general theory; therefore I have included alternative (direct) proofs of the important results. With careful choice of material, the book could thus form the basis for courses at different levels. The text is supplemented where appropriate by examples and exercises, the latter ranging greatly in degree of difficulty. Some are intended merely to fill in gaps in the text, for example caused by deliberately sketchy proofs of theorems; many of the others are intended to point the way to generalizations and extensions of the material covered, and most of these are accompanied by references. I have tried to use a notation which is, as far as possible, consistent throughout, standard, and which does not give rise to serious conflict. Because of the range of subject matter, there are times when attempting to maintain the first two of these objectives has placed some strain on the third, although I foresee no real difficulty.

I am grateful to Professor D. S. Jones, not only for his initial encouragement to undertake the writing of this book, but also for his comments on an early draft of several of the chapters. I would also like to thank Dr D. F. Griffiths for reading the manuscript and offering much constructive criticism. Finally I wish to acknowledge the expert typing of Mrs R. Hume, who typed most of the manuscript, and Mrs C. McLeod.

Dundee   **G. A. Watson**
August, 1979

1

Introduction

1.1 Approximation in a normed linear space

This book is concerned with the solution of problems in approximation theory. To motivate the study, and to give the flavour of the kind of problems which arise, we begin by giving some simple examples. One important area in which approximation is required is in the analysis of data. Suppose that, in an experimental situation, a set of data is generated which takes the form of measurements at m discrete time intervals. Suppose further that, from theoretical considerations, it is known (or suspected) that the measured data depend continuously on the time t, and can indeed be modelled by a function $f(a_1, a_2, \ldots, a_n, t)$ whose form is known, but for which the values of the parameters $a_1, a_2, \ldots, a_n$ are unknown. For example f may be a polynomial of degree $n - 1$ in t, or a sum of terms of the form $a_i e^{a_{i+1} t}$. Generally, it will be the case that there are more measurements available than unknowns, and also that these measurements are subject to experimental errors, so that the given model will not fit the data exactly for any choice of parameters. A typical approximation problem is therefore that of determining a set of parameter values which gives a 'best' fit to the experimental data.

A second example arises from the requirement of numerical analysts to have available computer software for elementary and special functions. Such functions can be provided in a number of ways, one of which is to represent them by simpler functions. For example, the tangent function $\tan(\pi x/4)$ may be represented conveniently in the range $-1 \leqslant x \leqslant 1$ by an expression of the form

$$x \sum_{i=1}^{n} a_i \phi_i(x)$$

where $\phi_i(x)$ is a certain polynomial of degree $2i$. The task of choosing appropriate values of the coefficients a_i can be regarded as an approximation problem.

As a final example, consider the operator equation

$$L(y(x)) = 0, \tag{1.1}$$

where L is perhaps a differential operator, so that (1.1) is just an ordinary differential equation in $y(x)$, with appropriate additional conditions to ensure existence and uniqueness of y. If an exact solution y is not available (or readily and conveniently obtained), then an approximate solution may be obtained by representing $y(x)$ by a function of x which also contains a number of free parameters. These parameters may be adjusted in a particular way in order to obtain a representation which is good in some well-defined sense; again this may be done by setting up and solving an approximation problem.

The foregoing examples may be interpreted as special cases of the following abstract problem: given a set S, a set M which is a subset of S, and a point $g \in S$, find a point of M nearest to g. In the first example, g is a vector in m-space whose components are the measurements; M is the subspace of vectors whose components are $f(a_1, a_2, \ldots, a_n, t)$ for each value of t. In the other two examples, S may be in the space of continuous functions, with $g = \tan(\pi x/4)$ and $g = 0$ respectively. The idea of a nearest point clearly implies some measure of distance between elements of the set S, and this suggests that an appropriate general setting for approximation problems is a metric space. Of particular importance in this book is a particular form of metric space appropriate to sets S which are linear, and on whose elements is defined a norm.

Definition 1.1 Let S be a linear space, and let $\| \, . \, \|$ be a real-valued function defined on the elements of S such that

    (i) $\| x \| > 0$ unless $x = 0$

    (ii) $\| \lambda x \| = | \lambda | \, \| x \|$ where λ is a scalar

    (iii) $\| x + y \| \leqslant \| x \| + \| y \|$.

    Then $\| \, . \, \|$ defines a *norm* on S.

Definition 1.2 A linear space S equipped with a norm is called a *normed linear space*.

In a normed linear space, a metric is defined by $\| x - y \|$, where x and y are elements of the space. Some examples of normed linear spaces are as follows.

Example The m-dimensional space R^m of real column vectors $\mathbf{x}$ with the L_p norm

$$\| \mathbf{x} \|_p = \left(\sum_{i=1}^{m} |x_i|^p \right)^{1/p} \qquad 1 \leqslant p < \infty$$

$$\| \mathbf{x} \|_\infty = \max_i |x_i|$$

where x_i denotes the ith component of $\mathbf{x}$. The case $p = 2$ is the *Euclidean* norm; the case $p = \infty$ is the *Chebyshev* (or minimax) norm.

Example The space $C[a, b]$ of continuous real-valued functions defined on the interval $[a, b]$ with the L_p norm

$$\|f\|_p = \left[\int_a^b |f(x)|^p dx \right]^{1/p} \qquad 1 \leqslant p < \infty$$

$$\|f\|_\infty = \max_{a \leqslant x \leqslant b} |f(x)|.$$

Definition 1.3 Given a point g and a set M in a normed linear space S, a point of M of minimum distance from g is called a *best approximation*, and the problem of determining such a point a *best approximation problem*.

In many practical approximation problems, while the space in which the problem is set is clear, the particular norm to be used is not, and indeed the choice of norm may form an important part of the whole analysis. We will not deal with this aspect of the problem to any great extent, but will concentrate rather on the treatment of problems for which the normed linear space S, set M, and point g are assumed given; motivation for the use of any particular norm will, however, be given where appropriate. Our main aims are then to investigate important examples of such problems, with particular regard to

  (i) the existence of best approximations
  (ii) the uniqueness of best approximations
  (iii) the characterization of best approximations
  (iv) the construction of methods for determining best approximations.
It is clear that the *existence* of a solution to an approximation problem requires certain additional conditions to be satisfied. As a trivial example, let S be the real line and M be the open interval $(0, 1)$. Then the problem of determining the point of M nearest to -1, say, has no solution. For a less trivial example, let $S = R^3$, $g = \mathbf{0}$ and let M be the subspace of vectors of the form

$$\mathbf{f} = (e^{a_1} + a_2, a_2/a_1^2, a_2^3)^T$$

where a_1 and a_2 are free parameters. (The superscript T will be used to denote transposition, and $\mathbf{0}$ will be used to denote a column vector whose elements are all zero.) If a_2 is set to zero, then for any norm $\|\mathbf{f}\|$ can be made arbitrarily small by letting a_1 become arbitrarily large in a negative direction. Thus at a solution we must have $e^{a_1} = 0$, which has no finite solution. The *uniqueness* of a best approximation also cannot in general be guaranteed. For example, in R^2 let M be the line $x_1 = 1$ and let g be the origin. Then if $\|\mathbf{x}\| = \max\{|x_1|, |x_2|\}$, any point $(1, x_2)^T$ such that $-1 \leqslant x_2 \leqslant 1$ is a point of minimum distance from g.

The task of *characterizing* best approximations is that of deriving potentially useful conditions which a best approximation must satisfy, and which, if satisfied, guarantee that a particular approximation is indeed a best approximation. A number of important approximation problems have solutions which are characterized by conditions which can be used *directly* as a means of actually calculating such best approximations. The construction of these solutions by these (or by other) means forms an important part of this book, and many of the theoretical results are included because of their value in providing motivation for numerical procedures or algorithms. The aims (i)–(iii) listed above can be usefully treated in some generality, and such general results are given in the remainder of the chapter. Of primary interest is the fundamental question of existence of a best approximation.

Exercises

1. Show that, in R^m, $\lim\limits_{p \to \infty} \|z\|_p = \max\limits_i |z_i|$ and in $C[a, b]$, $\lim\limits_{p \to \infty} \|f\|_p$ $= \max\limits_{a \leqslant x \leqslant b} |f(x)|$.

2. Determine which of the following define norms
    (i) in R^2: $\|z\| = \max\{3|z_1| + 2|z_2|, 2|z_1| + 3|z_2|\}$
    (ii) in R^m: $\|a\| = \max\limits_{0 \leqslant x \leqslant 1} \left| \sum\limits_{k=1}^{m} a_k x^{k-1} \right|$
    (iii) in the space $C^1[a, b]$ of continuously differentiable functions defined on $[a, b]$:

$$\|f\| = \max\limits_{a \leqslant x \leqslant b} [|f(x)|, |Df(x)|]$$

where D denotes differentiation with respect to x.

    (iv) in $C^1[a, b]$: $\|f\| = \max\limits_{a \leqslant x \leqslant b} |Df(x)|$

    (v) in R^m: $\|z\| = \left(\sum\limits_{i=1}^{m} |z_i|^p \right)^{1/p}$ $0 < p < 1$

    (vi) in R^m: $\|z\| = \sup\limits_{\|y\|_p \leqslant 1} y^T z$ $1 \leqslant p \leqslant \infty$.

3. If $r, s \in R^m$ prove that
    (i) $|r^T s| \leqslant \|r\|_2 \|s\|_2$ (Cauchy–Schwartz inequality)
    (ii) $|r^T s| \leqslant \|r\|_1 \|s\|_\infty$.

4. If $r \in R^m$ prove that
    (i) $\|r\|_p \leqslant \|r\|_\infty m^{1/p}$
    (ii) $\|r\|_p \geqslant \|r\|_\infty$
    for $1 \leqslant p < \infty$.

5. Let

$$\frac{1}{p} + \frac{1}{q} = 1$$

where $p \geqslant 1$. Prove that for $x > 0$,

$$x^{1/p} - \frac{x}{p} \leqslant \frac{1}{q}.$$

By setting $x_i = |r_i|^p |s_i|^{-q}, i = 1, 2, \ldots, m$, prove the generalization of exercise 3:

$$|\mathbf{r}^T\mathbf{s}| \leqslant \|\mathbf{r}\|_p \|\mathbf{s}\|_q \qquad \text{(Holder's inequality)}.$$

Deduce that

$$\|\mathbf{r}\|_p = \max_{\|\mathbf{s}\|_q \leqslant 1} \mathbf{r}^T\mathbf{s}.$$

6. If $\mathbf{r} \in R^m$, prove that

(i) $\|\mathbf{r}\|_p \geqslant \|\mathbf{r}\|_1 m^{(1/p - 1)}$

(ii) $\|\mathbf{r}\|_p \leqslant \|\mathbf{r}\|_1$

for $1 < p \leqslant \infty$. (Hint: use Holder's inequality.)

7. In $C[a, b]$ derive the Cauchy–Schwartz inequality

$$\left| \int_a^b f(x)g(x)\,dx \right| \leqslant \|f(x)\|_2 \|g(x)\|_2.$$

Under what conditions is equality attained?

8. If $f(x) \in C[a, b]$ satisfies the Lipschitz condition

$$|f(x) - f(y)| \leqslant L|x - y| \qquad a \leqslant x, y \leqslant b,$$

then

(i) $\|f\|_p \leqslant (b - a)^{1/p} \|f\|_\infty$

(ii) $\|f\|_p \geqslant \left(\frac{\|f\|_\infty}{Q(p+1)} \right)^{1/p} \|f\|_\infty$

where $Q = \max\{L, 2\|f\|_\infty/(b - a)\}$ (Hebden, 1971).

9. Let $f(x)$ be as in exercise 8. The techniques required there may be used to show that

(i) $\|f\|_p \geqslant (b - a)^{(1/p - 1)} \|f\|_1$

(ii) $\|f\|_p \leqslant \left(\frac{2Q}{\|f\|_\infty} \right)^{1 - 1/p} \|f\|_1$

where Q is as before.

1.2 Existence of best approximations

From the examples of non-existence given previously, it is suggested that closure of the subspace M of approximations is the required additional assumption. In fact, the rather stronger condition of compactness, so that *all* sequences in M have convergent subsequences, turns out to be the minimal property which will guarantee that a best approximation exists.

Theorem 1.1   Let M denote a compact set in a normed linear space. Then to each point g of the space there corresponds a point of M closest to g.

Proof   Let $\delta = \inf\{\|g - x\|, x \in M\}$. It follows from the definition of inf that there exists a sequence of points $x_1, x_2, \ldots$ in M such that

$$\|g - x_n\| \to \delta \qquad \text{as } n \to \infty.$$

Since M is compact, it follows that there exists a subsequence of $x_1, x_2, \ldots$ (which we will not rename) converging to $x^* \in M$. Now

$$\|g - x^*\| \leqslant \|g - x_n\| + \|x_n - x^*\|$$

and so, letting $n \to \infty$, $\|g - x^*\| \leqslant \delta$ since the left hand side is independent of n. However, since $x^* \in M$,

$$\|g - x^*\| \geqslant \delta.$$

Thus $\|g - x^*\| = \delta$ and x^* is a point of minimum distance from g in M. $\square$

While compactness of M is a sufficient condition for a best approximation to exist, it is clearly not necessary. For, consider the subset of the real line defined by the interval $(-\infty, 1]$. This is not compact, yet the nearest point to any g exists. It is often possible to restrict the set of approximations to a compact *subset* of M, and thereby obtain the required result: for any $g \geqslant \frac{1}{2}$, we could choose the best approximation from $[\frac{1}{2}, 1]$.

In order to make use of Theorem 1.1 to give particular results, it is necessary to obtain conditions on M which lead to compactness. We require the following definitions.

Definition 1.4   A subset of a normed linear space is said to be *bounded* if it is contained in some set of the form

$$\{x : \|x\| \leqslant C\}$$

where C is a constant.

Definition 1.5   A mapping T from a normed linear space S_A, into another normed linear space S_B, is *continuous* if $\|x_i - x\|_A \to 0$ implies that $\|Tx_i - Tx\|_B \to 0$ where x and the members of the sequence $\{x_i\}$ are elements of S_A, and $\|\cdot\|_A$, $\|\cdot\|_B$ denote norms on S_A and S_B respectively.

Our main existence result depends on the following lemma.

Lemma 1.1 Every closed, bounded, finite dimensional set in a normed linear space is compact.

Proof Let M be such a set, with dimension n. Then there exists a linearly independent set $\{m_1, m_2, \ldots, m_n\}$ such that each element of M may be written uniquely as

$$m = \sum_{i=1}^{n} \lambda_i m_i.$$

Let R^n be equipped with the L_∞ norm, and let T denote the mapping from $\lambda \in R^n$ into $m \in M$. Then

$$\|T\lambda - T\mu\| = \left\| \sum_{i=1}^{n} \lambda_i m_i - \sum_{i=1}^{n} \mu_i m_i \right\|$$

$$= \left\| \sum_{i=1}^{n} (\lambda_i - \mu_i) m_i \right\|$$

$$\leqslant \sum_{i=1}^{n} |\lambda_i - \mu_i| \, \|m_i\|$$

$$\leqslant \|\lambda - \mu\|_\infty \sum_{i=1}^{n} \|m_i\|.$$

Thus T is continuous. Now let $X = \{\lambda : T\lambda \in M\}$. The compactness of M will follow from that of X, since a continuous mapping from one metric space into another preserves compactness (exercise 12). We require, therefore, to show that X is closed and bounded. To show that X is closed, let $\lambda^{(k)} \in X$, $\lambda^{(k)} \to \lambda$. Then by continuity

$$T\lambda = T\left(\lim_{k \to \infty} \lambda^{(k)} \right) = \lim_{k \to \infty} T(\lambda^{(k)}).$$

Since M is closed, $T\lambda \in M$ and so $\lambda \in X$. Thus X is closed. Now the set $\{\lambda : \|\lambda\|_\infty = 1\}$ is compact, and T is continuous so

$$\alpha = \inf_{\|\lambda\|_\infty = 1} \|T\lambda\| = \inf_{\|\lambda\|_\infty = 1} \left\| \sum_{i=1}^{n} \lambda_i m_i \right\|$$

is attained (exercise 11). Further, since the functions $\{m_i\}$ are linearly independent, $\alpha > 0$. Thus for any $\lambda \neq \mathbf{0}$,

$$\|T\lambda\| = \|T(\lambda/\|\lambda\|_\infty)\|\lambda\|_\infty\| = \|T(\lambda/\|\lambda\|_\infty)\| \, \|\lambda\|_\infty$$

$$\geqslant \alpha \|\lambda\|_\infty.$$

Now $\|T\lambda\|$ is bounded for $\lambda \in X$ since M is, and so $\|\lambda\|_\infty$ is bounded for $\lambda \in X$. Thus X is bounded. $\qquad\square$

Theorem 1.2 Let M be a finite dimensional subspace of a normed linear space S. Then there exists a best approximation in M to any point of S.

Proof Let M be such a subspace, and let g be the prescribed point. Then if $\bar{m}$ is an arbitrary point of M, the point sought lies in the set

$$\{m : m \in M, \|m - g\| \leqslant \|\bar{m} - g\|\}. \tag{1.2}$$

This set is closed and bounded and thus compact by Lemma 1.1. The result follows from Theorem 1.1. $\qquad\square$

It is not possible to drop the finite dimensional requirement of this theorem. For, let S be the space of continuous functions defined on $[0, \frac{1}{2}]$ with the L_∞ norm, let M be the subspace of polynomials of any degree, and let $g = 1/(1 - x)$. Then given $\varepsilon > 0$, there exists N such that

$$\left| g - (1 + x + x^2 + \ldots + x^N) \right| < \varepsilon \qquad 0 \leqslant x \leqslant \tfrac{1}{2}.$$

Thus any best approximation, say p^*, would have to satisfy

$$\|g - p^*\| = 0$$

which implies $p^* = 1/(1 - x)$. This is clearly impossible, and so no best approximation exists.

Exercises

10. Verify that the following are not compact
   (i) R^m
   (ii) $C[0, 1]$ (Let $f_n(x) = x^n$.)
   (iii) The space l_2 of infinite numerical sequences.

11. Prove that a continuous real-valued function defined on a compact subset of a normed linear space attains its minimum and maximum values on that set.

12. Prove that a continuous mapping from one normed linear space into another preserves compactness of sets.

13. Using first principles, and elementary analysis, solve the following approximation problems.
   (i) Find the best approximation to
      (a) e^x on $[0, 1]$
      (b) $f(x) = 0$, $\varepsilon \leqslant |x| \leqslant 1$, $f(x) = 1 - |x|/\varepsilon$, $|x| < \varepsilon$, on $[-1, 1]$
      by a constant in the L_1, L_2, and L_∞ norms.
   (ii) Find a to minimize $\displaystyle \max_{0 \leqslant x \leqslant 1} |x^4 - ax|$
   (iii) Find a to minimize $\int_0^1 |x - ax^2| \, dx$.

14. Let

$$\mathbf{r}(\mathbf{a}) = \mathbf{b} - A\mathbf{a}$$

where $\mathbf{a} \in R^n$, A is a constant $m \times n$ matrix, and $\mathbf{b}$ a constant vector in R^m. If $\mathbf{a}^{(p)}$

minimizes $\|\mathbf{r(a)}\|_p$ prove that

$$\lim_{p \to \infty} \left\| \mathbf{r}^{(p)} \right\|_p = \left\| \mathbf{r}^{(\infty)} \right\|_\infty$$

where

$$\mathbf{r}^{(p)} = \mathbf{b} - A\mathbf{a}^{(p)} \qquad 1 \leqslant p \leqslant \infty.$$

(Hint: Use the results of exercise 4 to obtain an appropriate string of inequalities, and let $p \to \infty$).

1.3 Uniqueness of best approximations

Theorem (1.2) shows that the search for a best approximation to g from M may be confined to sets of the form

$$\{m : m \in M, \|m - g\| \leqslant r\}, \tag{1.3}$$

where r is a constant. For example, in the space R^2 equipped with the L_2 norm, the set $\|m - g\| \leqslant r$ just defines the interior and boundary of a circle, centre g and radius r, and, geometrically, we are seeking the circle of minimum radius which passes through a point of M (see Figure 1).

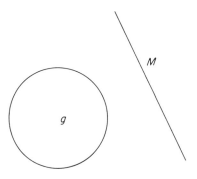

Figure 1

In general, the sets $\|m - g\| \leqslant r$ will be closed regions of different shapes, and the precise nature of the boundary would appear to play an important role in the question of uniqueness. This is in fact the case, and we begin the study of uniqueness by a closer examination of the properties of these closed regions.

Definition 1.6　In a normed linear space S, the set of elements $x \in S$ satisfying

$$\|x - a\| \leqslant r \tag{1.4}$$

for given $a \in S$, constant r, is called a *closed sphere* with centre a and radius r.
A fundamental property possessed by these closed spheres is that of convexity.

Definition 1.7   A set M of a linear space S is *convex* if x, $y \in M$ implies that $\lambda x + (1 - \lambda) y \in M$ for all λ satisfying $0 \leqslant \lambda \leqslant 1$.

Geometrically, a set is convex if all line segments joining pairs of points in the set also belong to the set. Thus, of the examples in R^2 shown in Figure 2, (i) and (ii) are convex, while (iii) is not.

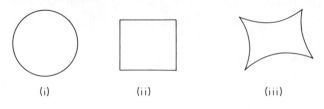

(i)　　　　　　　(ii)　　　　　　　(iii)

Figure 2

Lemma 1.2   Closed spheres are convex.

Proof   Let K be the set of elements

$$K = \{ x : x \in S, \ \| x - a \| \leqslant r \}.$$

Let x_1, $x_2 \in K$ and let λ satisfy $0 \leqslant \lambda \leqslant 1$. Then

$$\| \lambda x_1 + (1 - \lambda) x_2 - a \| \leqslant \lambda \| x_1 - a \| + (1 - \lambda) \| x_2 - a \|$$
$$\leqslant r. \qquad \square$$

The proof of this lemma depends on the use of the triangle inequality for norms (inequality (iii) in Definition 1.1). In fact, the convexity of the closed spheres is a direct consequence of the associated norm being a convex function on the elements of S.

Definition 1.8   Let $f(x)$ be a function on the elements x of a linear space S. Then $f(x)$ is a *convex function* if

$$f(\lambda x_1 + (1 - \lambda) x_2) \leqslant \lambda f(x_1) + (1 - \lambda) f(x_2) \tag{1.5}$$

for any x_1, $x_2 \in S$ and any λ satisfying $0 \leqslant \lambda \leqslant 1$. If strict inequality holds in (1.5) when $0 < \lambda < 1$, $f(x)$ is a *strictly convex function*.

It is easily seen that if S is a normed linear space, then the norm is a convex function on the elements of S. The norm can not of course be a strictly convex function (set $x_1 = 0$); however we say that S is a *strictly convex normed linear space* if, when elements x_1 and x_2 lie on the boundary of the closed sphere (1.4), then it follows that

$$\| \lambda x_1 + (1 - \lambda) x_2 - a \| < r \qquad 0 < \lambda < 1. \tag{1.6}$$

Geometrically, the boundary of the closed sphere contains no line segments, so that any line joining points on the boundary lies completely (with the exception of the two end points) within the sphere. Intuitively (see Figure 1) this would appear to be the appropriate requirement for uniqueness of the best approximation, and the following theorem shows that this is indeed the case.

Theorem 1.3 In a strictly convex normed linear space S, a finite dimensional subspace M contains a unique best approximation to any point $g \in S$.

Proof The existence of at least one best approximation follows from Theorem 1.2. Let m_1 and m_2 be elements of M of minimum distance r from g. Then for $0 < \lambda < 1$,

$$\| \lambda m_1 + (1 - \lambda)m_2 - g \| \leqslant \| \lambda (m_1 - g) \| + \| (1 - \lambda)(m_2 - g) \|$$
$$\leqslant r$$

so that

$$\| m_1 - g \| = \| m_2 - g \| = \| \lambda m_1 + (1 - \lambda)m_2 - g \|.$$

This contradicts the inequality (1.6) unless $m_1 = m_2$. $\square$

Theorem 1.3 shows that the uniqueness of the best approximation will be an immediate consequence of the strict convexity of the appropriate closed spheres. In particular, this information will be obtained by an examination of the *unit balls* (obtained by setting $a = 0$, $r = 1$). It is convenient to illustrate the case of the space R^2, equipped with the different L_p norms, $1 \leqslant p \leqslant \infty$, and some of the unit balls are shown in Figure 3.

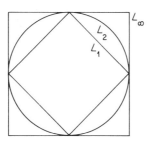

Figure 3

The illustrations show that R^2 with the L_1 or L_∞ norms is not a strictly convex normed linear space, and suggest that R^2 with the L_p norms, $1 < p < \infty$, is. This is indeed the case, and in fact these properties of the unit balls carry over to the spaces R^n and $C[X]$ (the space of continuous real-valued functions defined on X, where X is for example an N-dimensional continuum) with the L_p norms, $1 \leqslant p \leqslant \infty$.

Exercises

15. Let $f(\mathbf{x}):R^n \to R$ be a continuously differentiable mapping (where $R \equiv R^1$, the real line), and let $f(\mathbf{x})$ be convex. Then if $\mathbf{d}\in R^n$ has ith component $d_i = \partial f(\mathbf{x})/\partial x_i$, show that for any $\mathbf{x}, \mathbf{y}\in R^n$

$$f(\mathbf{y}) \geqslant f(\mathbf{x}) + \mathbf{d}^T(\mathbf{y}-\mathbf{x}).$$

Is the converse true?

16. Let $f(\mathbf{x}):R^n \to R$ be defined by

$$f(\mathbf{x}) = \mathbf{x}^T A \mathbf{x} + \mathbf{b}^T \mathbf{x} + c,$$

where A is a constant $n \times n$ symmetric matrix, and $\mathbf{b}\in R^n$ and $c\in R$ are constant. Show that if A is positive semi-definite, then $f(\mathbf{x})$ is convex. Is the converse true?

17. Prove that R^n with the L_2 norm is a strictly convex normed linear space. Prove the same result for $C[a, b]$. Try to prove (or look up) the corresponding results for the other L_p norms, $1 < p < \infty$ (see Hardy, Littlewood, and Polya, 1934).

18. Illustrate diagrammatically for the L_1 and L_∞ norms in R^2 the approximation problems analogous to that depicted in Figure 1, where M is a straight line. When are the best approximations in these norms unique?

1.4 Some additional results concerning convexity

The property of convexity plays an important role in questions other than just those of uniqueness, and it is convenient to include some other results at this point. In particular, corresponding to any set D in R^n, there exists a unique convex set called the convex hull of D, denoted by $\operatorname{conv}(D)$.

Definition 1.9 Let D be any set in R^n. Then the convex hull of D, $\operatorname{conv}(D)$, is the set of points which are expressible as convex linear combinations of the elements of D, that is as finite sums of the form $\sum_i \lambda_i \mathbf{d}_i$ with $\mathbf{d}_i \in D$, $\lambda_i \geqslant 0$, for all i, and $\sum_i \lambda_i = 1$.

Theorem 1.4 (Carathéodory's theorem) Let $D \subset R^n$. Then any point of $\operatorname{conv}(D)$ can be expressed as a convex linear combination of $(n+1)$ or fewer elements of D.

Proof Let $\mathbf{k}\in\operatorname{conv}(D)$ be such that

$$\mathbf{k} = \sum_{i=0}^{m} \lambda_i \mathbf{d}_i$$

with $\lambda_i > 0$, $\sum\limits_{i=0}^{m} \lambda_i = 1$, $\mathbf{d}_i \in D$. Then the vectors $\mathbf{k}_i = \mathbf{d}_i - \mathbf{k}$ satisfy

$$\sum_{i=0}^{m} \lambda_i \mathbf{k}_i = \mathbf{0}.$$

If $m \leqslant n$ we have finished. Otherwise we must have

$$\sum_{i=1}^{m} \mu_i \mathbf{k}_i = \mathbf{0}$$

for some coefficients μ_i, not all zero. Thus, defining $\mu_0 = 0$, for all α

$$\sum_{i=0}^{m} (\lambda_i + \alpha \mu_i) \mathbf{k}_i = \mathbf{0}.$$

Now let $|\alpha|$ be increased away from zero until one of the coefficients $\lambda_i + \alpha \mu_i$ vanishes. The remaining coefficients stay non-negative and cannot all vanish since $\lambda_0 > 0$. Now replace $\mathbf{k}_i$ by $\mathbf{d}_i - \mathbf{k}$ giving

$$\mathbf{k} \sum_{i=0}^{m} (\lambda_i + \alpha \mu_i) = \sum_{i=0}^{m} (\lambda_i + \alpha \mu_i) \mathbf{d}_i.$$

Dividing through by the coefficient of $\mathbf{k}$ on the left hand side, we obtain an expression for $\mathbf{k}$ which contains fewer than $(m+1)$ terms. This process can be repeated until $m \leqslant n$. $\qquad\square$

Theorem 1.5 Let D be a closed, convex subset of R^n. Then D does not contain the origin if and only if there exists $\mathbf{z} \in R^n$ such that

$$\mathbf{d}^T \mathbf{z} > 0 \qquad \text{for all } \mathbf{d} \in D.$$

Proof Assume $\mathbf{0} \notin D$ and consider the problem:

$$\text{find } \mathbf{d} \in D \text{ to minimize } \|\mathbf{d}\|_2.$$

We may seek approximations from the set

$$\{\mathbf{d} \in D : \|\mathbf{d}\|_2 \leqslant \|\bar{\mathbf{d}}\|_2\}$$

where $\bar{\mathbf{d}} \in D$ is arbitrary, and since this set is compact, the existence of a point $\mathbf{z} \in D$ at which the minimum is attained is guaranteed. Now if $\mathbf{d} \in D$ is arbitrary, by the convexity of D

$$\lambda \mathbf{d} + (1 - \lambda) \mathbf{z} \in D \qquad 0 \leqslant \lambda \leqslant 1.$$

Thus

$$0 \leqslant \|\lambda \mathbf{d} + (1 - \lambda)\mathbf{z}\|_2^2 - \|\mathbf{z}\|_2^2$$
$$= \lambda^2 \|\mathbf{d} - \mathbf{z}\|_2^2 + 2\lambda(\mathbf{d} - \mathbf{z})^T \mathbf{z}.$$

This inequality cannot be valid for small positive λ unless

$$(\mathbf{d} - \mathbf{z})^T \mathbf{z} \geq 0$$

and so

$$\mathbf{d}^T \mathbf{z} \geq \mathbf{z}^T \mathbf{z} > 0,$$

which, since $\mathbf{d}$ is arbitrary, gives the required conclusion.

Now assume that there exists $\mathbf{z} \in R^n$ such that

$$\mathbf{d}^T \mathbf{z} > 0 \qquad \text{for all } \mathbf{d} \in D.$$

If D contains the origin, this is clearly impossible. $\qquad\qquad\qquad\square$

Exercises

19. Prove that a closed, convex subset of R^n possesses a unique point of minimum L_2 norm.

20. Let $f(\mathbf{x})$ be a convex function, and let $\mathbf{x}^*$ satisfy

$$f(\mathbf{x}^*) \leq f(\mathbf{x})$$

for all $\mathbf{x}$ in a neighbourhood of $\mathbf{x}^*$ (i.e. $\mathbf{x}^*$ is a local minimum). Prove that $f(\mathbf{x}^*) \leq f(\mathbf{x})$ for all $\mathbf{x}$.

21. Prove that the intersection of m compact convex sets in R^n is non-empty if and only if every $(n + 1)$ of the sets has a non-empty intersection. (Helly's theorem; see for example Rademacher and Schoenberg, 1950.)

22. A point $\mathbf{x}$ is an *extreme point* or *vertex* of a convex set in R^n if and only if there do not exist other points $\mathbf{x}_1$, $\mathbf{x}_2$, $\mathbf{x}_1 \neq \mathbf{x}_2$, of the set such that for some λ, $0 < \lambda < 1$,

$$\mathbf{x} = \lambda \mathbf{x}_2 + (1 - \lambda)\mathbf{x}_1.$$

Let $f(\mathbf{x}) = \mathbf{a}^T \mathbf{x}$, where $\mathbf{a} \in R^n$. Show that the minimum of $f(\mathbf{x})$ on a closed, convex set is attained at an extreme point.

23. Let K be a closed, convex *cone* in R^n (i.e. $\mathbf{k} \in K$ implies that $\alpha \mathbf{k} \in K$, $\alpha \geq 0$) and B a closed, bounded, convex set in R^n. Show that the intersection of K and B is empty if and only if there exists $\mathbf{d} \in R^n$ such that

$$\begin{aligned} \mathbf{d}^T \mathbf{k} &\leq 0 \qquad \text{for all } \mathbf{k} \in K, \\ \mathbf{d}^T \mathbf{b} &> 0 \qquad \text{for all } \mathbf{b} \in B. \end{aligned}$$

1.5 Characterization of best approximations

In this section, we give a general characterization result for an important class of linear approximation problems, which includes all problems which may be expected to arise in practice, and all those which are dealt with in detail later on.

The result is in fact obtained by specializing from a wider class of nonlinear problems, for which necessary conditions for a best approximation are obtained. The unified nature of such results is therefore demonstrated, and particular theorems required in subsequent chapters will be shown to be easily obtained as special cases of this theory. (For a completely general treatment of the linear case, see Singer, 1970.) The analysis required here of necessity draws on certain concepts and results from functional analysis, and readers without the necessary background may omit this section.

The basic requirement here is a (possibly nonlinear) mapping $f(\mathbf{a})$ from R^n into a subset M of an arbitrary normed linear space S. Then, without loss of generality, the problem of finding a best approximation from M to an element of S can be stated

$$\text{find } \mathbf{a} \in R^n \text{ to minimize } \|f(\mathbf{a})\|. \tag{1.7}$$

Let S^* be the dual space of S, that is the space of continuous linear functionals $v(f)$ defined on S. For convenience, we will write

$$v(f) = \langle f, v \rangle,$$

thus defining the linear functional as an inner product between the elements of S and those of S^*. The dual norm on S^* can then be written

$$\|v\|^* = \sup_{\|f\| \leqslant 1} \langle f, v \rangle.$$

Now let the set $V(f)$ be defined by

$$V(f) = \{v \in S^* : \|f\| = \langle f, v \rangle, \|v\|^* \leqslant 1\}. \tag{1.8}$$

Then, assuming that the partial derivative of f with respect to a_j exists and is denoted by $g_j(\mathbf{a}), j = 1, 2, \ldots, n$, we can give the following necessary condition for $\mathbf{a}$ to solve (1.7).

Theorem 1.6 Let $\mathbf{a}$ solve (1.7) and assume that there exists a neighbourhood of $\mathbf{a}$ in which we can write

$$f(\mathbf{a} + \mathbf{z}) = f(\mathbf{a}) + \sum_{j=1}^{n} z_j g_j(\mathbf{a}) + o(\|\mathbf{z}\|_n) \tag{1.9}$$

where $\|.\|_n$ denotes any norm on R^n. Then there exists $v \in V(f(\mathbf{a}))$ such that

$$\langle g_j, v \rangle = 0 \qquad j = 1, 2, \ldots, n. \tag{1.10}$$

Proof Let $\mathbf{a}$ solve (1.7) and let (1.9) be satisfied but (1.10) not be. Then since V is a convex set, so is $D \in R^n$ defined by

$$D = \{\mathbf{d} : d_j = \langle g_j, v \rangle, j = 1, 2, \ldots, n, v \in V(f(\mathbf{a}))\}$$

and so, by Theorem 1.5, there exists $\mathbf{z} \in R^n$ such that

$$\mathbf{d}^T \mathbf{z} < 0 \qquad \text{for all } \mathbf{d} \in D,$$

or

$$\sum_{j=1}^{n} z_j \langle g_j, v \rangle \leqslant -\delta \qquad \text{for all } v \in V(f(\mathbf{a}))$$

for some $\delta > 0$. Now for any $v(\gamma) \in V(f(\mathbf{a} + \gamma \mathbf{z}))$ with $\gamma > 0$,

$$\|f(\mathbf{a} + \gamma \mathbf{z})\| = \langle f(\mathbf{a} + \gamma \mathbf{z}, v(\gamma) \rangle$$

$$= \langle f(\mathbf{a}), v(\gamma) \rangle + \gamma \sum_{j=1}^{n} z_j \langle g_j, v(\gamma) \rangle + O(\gamma)$$

$$< \langle f(\mathbf{a}), v(\gamma) \rangle - \gamma \delta + \gamma \sum_{j=1}^{n} z_j \langle g_j, v(\gamma) - v \rangle + O(\gamma)$$

$$(1.11)$$

for all $v \in V(f(\mathbf{a}))$. By the weak* compactness of the unit ball in S^* (Alaoglu–Bourbaki theorem, for example Holmes, 1975) there exists a positive sequence $\{\gamma_j\} \to 0$ and $v^* \in S^*$ such that

$$\langle u, v(\gamma_j) - v^* \rangle \to 0 \qquad \text{as } j \to \infty$$

for all $u \in S$. Further,

$$0 \leqslant \|f(\mathbf{a})\| - \langle f(\mathbf{a}), v(\gamma) \rangle$$
$$\leqslant \|f(\mathbf{a} + \gamma \mathbf{z})\| - \langle f(\mathbf{a}), v(\gamma) \rangle$$
$$= o(1)$$

and so $v^* \in V(f(\mathbf{a}))$. Letting $\gamma \to 0$ in the inequality (1.11) along the sequence $\{\gamma_j\}$ and setting $v = v^*$ we obtain a contradiction that $\mathbf{a}$ gives a minimum, and the result follows. $\qquad \square$

Theorem 1.7 Let $f(\mathbf{a})$ be a linear function of $\mathbf{a}$. Then $\mathbf{a}$ solves (1.7) if and only if there exists $v \in V(f(\mathbf{a}))$ such that

$$\langle g_j, v \rangle = 0 \qquad j = 1, 2, \ldots, n. \qquad (1.12)$$

Proof Necessity follows from Theorem 1.6. Let the conditions of this theorem be satisfied, and let $\mathbf{b}$ be any other vector in R^n. Then

$$f(\mathbf{b}) = f(\mathbf{a}) + \sum_{j=1}^{n} (b_j - a_j) g_j$$

(where g_j is of course independent of $\mathbf{a}$). Thus

$$\langle f(\mathbf{b}), v \rangle = \langle f(\mathbf{a}), v \rangle = \|f(\mathbf{a})\|$$

for $v \in V(f(\mathbf{a}))$ satisfying (1.12), and it follows that

$$\|f(\mathbf{a})\| \leqslant \|f(\mathbf{b})\|. \qquad \square$$

The derivation of particular characterization results from Theorem 1.7 of course requires that the form of inner product, and appropriate sets $V(f)$, be known. We will be almost exclusively concerned with approximation in L_p norms, $1 \leqslant p \leqslant \infty$, and the appropriate dual norm is therefore the L_q norm, where

$$\frac{1}{p} + \frac{1}{q} = 1.$$

For finite dimensional space R^m, the inner product is defined by

$$\langle \mathbf{f}, \mathbf{v} \rangle = \mathbf{f}^T \mathbf{v},$$

with $S^* = S = R^m$ and the appropriate sets $V(f)$ are defined as follows, where we assume that $\|\mathbf{f}\| \neq 0$:

$$L_1 : V(\mathbf{f}) = \{\mathbf{v} : |v_i| \leqslant 1,\ v_i = \text{sign}\,(f_i) \text{ if } f_i \neq 0\}$$

$$L_p : V(\mathbf{f}) = \{\mathbf{v} : v_i = f_i |f_i|^{p-2} / \|\mathbf{f}\|^{p-1}\} \qquad 1 < p < \infty$$

$$L_\infty : V(\mathbf{f}) = \text{conv}\,\{\mathbf{e}_i \,\text{sign}\,(f_i),\ \text{all } i \text{ with } |f_i| = \|\mathbf{f}\|\}$$

where $\mathbf{e}_i$ denotes the ith coordinate vector.

For the space $C[X]$, where X is a compact measurable subset of R^N, the inner product is defined by

$$\langle f, v \rangle = \int_X f v \, \mathrm{d}x$$

and the sets $V(f)$ given by (again for $\|f\| \neq 0$)

$$L_1 : V(f) = \{v(x) : |v(x)| \leqslant 1,\ v(x) = \text{sign}\,(f(x)),\ f(x) \neq 0,\ x \in X\}$$

$$L_p : V(f) = \{v(x) : v(x) = f(x) |f(x)|^{p-2} / \|f\|^{p-1},\ x \in X\} \qquad 1 < p < \infty$$

$$L_\infty : V(f) = \text{conv}\,\{\text{sign}\,(f(x))\delta(x),\ \text{all } x \in X \text{ with } |f(x)| = \|f\|\}$$

where $\delta(x)$ is the delta function given by

$$\langle f(x), \delta(\xi) \rangle = f(\xi).$$

Exercises

24. Prove that the sets defined by (1.8) are as stated for the L_p norms, and give the appropriate sets when $\mathbf{f} \in R^m$ and $f(x) \in C[X]$ are identically zero.

25. Let G be a symmetric positive definite $m \times m$ matrix, and let R^m be normed by the *elliptic norm*

$$\|\mathbf{f}\| = (\mathbf{f}^T G \mathbf{f})^{1/2}.$$

Determine the dual norm on R^m.

26. Let $F(f)$ be a convex function on the elements $f \in S$. Then the *subdifferential* (or set of subgradients) of $F(f)$ at f, denoted by $\partial F(f)$ is the set of elements $v \in S^*$

satisfying the subgradient inequality

$$F(m) \geqslant F(f) + \langle m - f, v \rangle \qquad \text{for all } m \in S.$$

Prove that $\partial F(f) \equiv V(f)$.

27.   Prove that if $\mathbf{f} \in R^m$ and $\| \mathbf{f} \|_p \neq 0$ then for $1 < p < \infty$ the set $V(\mathbf{f})$ contains a single vector which is just the vector of partial derivatives of $\| \mathbf{f} \|$ with respect to the components of $\mathbf{f}$. Extend this result to $C[X]$.

1.6 Some alternatives to best approximation

The main theme of this book is the solution of 'best approximation' problems: a subspace of possible approximations is defined, and the solution is required to be better (or at least no worse) than all other points of this subspace, in the sense of minimizing a particular norm. However, approximations may be sensibly defined in ways which are not encompassed by this formulation, and we digress here to review briefly some important examples in the space $C[a, b]$. In particular, we consider approximations to $f(x) \in C[a, b]$ by polynomials. These have long been regarded as good approximating functions, and one reason for this is that the sequence $\{ 1, x, x^2, \dots \}$ is fundamental in $[a, b]$. This is a consequence of the following classical theorem given by Weierstrass in 1885.

Theorem 1.8   Let $f(x) \in C[a, b]$. For any $\varepsilon > 0$, there exists a polynomial $P(x)$ such that

$$|f(x) - P(x)| < \varepsilon \qquad \text{for all } x \in [a, b].$$

This result may be proved in a number of ways; see for example Cheney (1966).
   Perhaps the simplest way in which a polynomial of degree n can be fitted to a given function $f(x)$ is by *interpolation* at prescribed points.

Theorem 1.9   There exists a unique polynomial of degree n which takes given values at $(n + 1)$ distinct points.

Proof   The matrix of the appropriate linear system is simply

$$
\begin{bmatrix}
1 & x_1 & x_1^2 & \cdots & x_1^n \\
1 & x_2 & x_2^2 & \cdots & x_2^n \\
 & & \vdots & \vdots & \\
1 & x_{n+1} & x_{n+1}^2 & \cdots & x_{n+1}^n
\end{bmatrix}
$$

where the points are $x_1, x_2, \dots, x_{n+1}$. The determinant of this matrix is Vandermonde's determinant, and has the value $\displaystyle \prod_{1 \leqslant i \leqslant j \leqslant n+1} (x_i - x_j)$ which is obviously nonzero if the points are distinct. $\square$

   Such an interpolating polynomial is often given explicitly in *Lagrange form*, when it is known as the Lagrange interpolation formula. Defining $(n+1)$ polynomials $l_i(x)$ of degree n by the interpolation conditions

$$l_i(x_j) = \delta_{ij} = \begin{cases} 0 & i \neq j \\ 1 & i = j \end{cases} \quad \text{(the Kronecker delta)}$$

then the degree n polynomial $P_n(x)$ given explicitly by

$$P_n(x) = \sum_{j=1}^{n+1} f_i l_i(x)$$

satisfies

$$P_n(x_i) = f_i \qquad i = 1, 2, \ldots, n+1.$$

The polynomials $l_i(x)$ are easily seen to be given by

$$l_i(x) = \prod_{\substack{j=1 \\ j \neq i}}^{n+1} \frac{x - x_j}{x_i - x_j} \qquad i = 1, 2, \ldots, n+1. \tag{1.13}$$

If the values f_i are such that

$$f_i = f(x_i) \qquad i = 1, 2, \ldots, n+1$$

and $f(x)$ has $(n+1)$ continuous derivatives, we have (exercise 27)

$$\| f(x) - P_n(x) \|_\infty \leq \frac{1}{(n+1)!} \| D^{n+1} f \|_\infty \left\| \prod_{i=1}^{n+1} (x - x_i) \right\|_\infty, \tag{1.14}$$

where $D^{n+1} f$ is the $(n+1)$st derivative of f with respect to x, and the interval $[a, b]$ on which the norm is defined may be thought of as the smallest interval containing the points. The upper bound on the error can thus be made small by suitable choice of the points x_i; in fact precise values of x_i which give a minimum value can be obtained, as will become clear in Chapter 5.

Approximating polynomials may also be produced which interpolate not only in function value, but also in derivatives at prescribed points. This form of interpolation is known as Hermite interpolation. For example, we have the following result.

Theorem 1.10 There exists a unique polynomial of degree $2n+1$ which has given values, and given first derivative values, at $(n+1)$ distinct points.

Proof Let

$$h_i(x) = [1 - 2(x - x_i)Dl_i(x_i)]l_i^2(x)$$
$$h_i'(x) = (x - x_i)l_i^2(x),$$

for $i = 1, 2, \ldots, n+1$, where $l_i(x)$ is defined by equation (1.13). Then $h_i(x)$ and

$h_i'(x)$ are polynomials of degree $2n + 1$ which satisfy

$$\left.\begin{array}{ll} h_i(x_j) = \delta_{ij} & h_i'(x_j) = 0 \\ Dh_i(x_j) = 0 & Dh_i'(x_j) = \delta_{ij} \end{array}\right\} i, j = 1, 2, \ldots, n+1.$$

as may readily be verified. Now let $P(x)$ be defined by

$$P(x) = \sum_{i=1}^{n+1} [f_i h_i(x) + f_i' h_i'(x)].$$

Then it is again readily verified that $P(x)$ can satisfy required conditions.

For uniqueness, let $Q(x)$ be any other such interpolating polynomial. Then $P(x) - Q(x)$ is a polynomial with zeros of multiplicity at least 2 at each point $x_i = 1, 2, \ldots, n+1$. But $P(x) - Q(x)$ is of degree $2n + 1$, and so must be identically zero. □

In the case where $f_i = f(x_i)$, $f_i' = Df(x_i)$, the error may be bounded as before. Provided that $f(x)$ has $(2n + 2)$ continuous derivatives, we have (exercise 28)

$$\|f(x) - P(x)\|_\infty \leq \frac{1}{(2n+2)!} \|D^{2n+2}f\|_\infty \left\| \prod_{i=1}^{n+1} (x - x_i)^2 \right\|_\infty. \qquad (1.15)$$

Analogous results hold for Hermite interpolation involving higher derivatives. A particular case of this is the classical Taylor polynomial, which we can consider as being the (unique) degree n polynomial which interpolates to $f(x)$ and derivatives to order n at the *single point* a. If $f(x)$ has $(n + 1)$ continuous derivatives, we have then

$$\|f(x) - P(x)\|_\infty \leq \frac{1}{(n+1)!} |x - a|^{n+1} \|D^{n+1}f\|_\infty. \qquad (1.16)$$

This particular form of approximation is useful only in a small neighbourhood of the point $x = a$.

The most general problem of approximating to function and derivative values may be given in the following form, which encompasses all the special cases given above. Let E be an $m \times (n + 1)$ matrix which contains elements E_{ij} all of which are zero except $(n + 1)$ which are equal to 1, and let a set of m points $x_1 < x_2 < \ldots < x_m$ be given. Then the problem of finding a polynomial $P(x)$ of degree n satisfying

$$E_{ij}(D^{j-1}P(x_i) - c_{ij}) = 0 \qquad i = 1, 2, \ldots, m; j = 1, 2, \ldots, n+1,$$

for some numbers c_{ij}, is a Hermite–Birkhoff (or sometimes just Birkhoff) interpolation problem. In this generality, it is clear that there may be no solution, and the problem of identifying appropriate matrices E which lead to a nonsingular system of equations has been extensively studied in recent years. See, for example, Karlin and Karon (1972), Lorentz (1974).

Weierstrass' theorem (Theorem 1.8) shows that there exists a sequence of polynomials of degree n, say $\{P_n(x)\}$ such that

$$\|f(x) - P_n(x)\|_\infty \to 0 \qquad \text{as } n \to \infty.$$

Unfortunately, this is not a result which holds in general when $P_n(x)$ is defined by interpolation. The following examples are classical.

Example (Runge, 1901) Let

$$f(x) = \frac{1}{1 + x^2}$$

and let $P_n(x)$ be defined by interpolation at $(n + 1)$ equispaced points in $[-5, 5]$. Then $\|f(x) - P_n(x)\|_\infty$ becomes *arbitrarily large* as $n \to \infty$.

Example (Bernstein, 1912) Let

$$f(x) = |x|$$

and let $P_n(x)$ be defined by interpolation at $(n + 1)$ equispaced points in $[-1, 1]$. Then $|f(x) - P_n(x)| \to 0$ *only* at the three points $x = -1, 0, 1$.

Suitable bunching of the interpolation points can improve matters in many cases. However, Faber (1914) has shown that for any given selection of points, it is possible to find a continuous function whose interpolating polynomial of degree n is such that

$$\|f(x) - P_n(x)\|_\infty \not\to 0$$

as $n \to \infty$.

These examples show that while interpolation is often useful, and can enable approximations in a convenient form to be obtained in a relatively simple manner, it must be used with some care. We will pick up the subject of interpolation again in Chapter 8. However, we conclude this chapter by looking briefly at a particular way of approximating by *rational functions*, that is, functions formed by the quotient of two polynomials. To motivate the particular criterion, we return to the Taylor polynomial: this may be defined through the solution to the following problem. Given $f(x)$ defined in $[-c, c]$, where $0 < c < 1$, determine the polynomial $P_n(x)$ of degree n which is such that

$$|f(x) - P_n(x)| \leqslant M|x^k| \qquad -c \leqslant x \leqslant c$$

for as large a value of k as possible, where M is a constant. Clearly if $f(x) \in C^{(n+1)}$ $[-c, c]$ we can take $k = n + 1$, and $P_n(x)$ to be the Taylor polynomial. Now, let $R(x)$ be a rational function of the form

$$R(x) = \frac{P_n(x)}{Q_m(x)} = \frac{\displaystyle\sum_{i=0}^{n} a_i x^i}{\displaystyle\sum_{j=0}^{m} b_j x^j},$$

and suppose we seek the coefficients so that

$$|f(x) - R(x)| \leq M|x^k| \qquad -c \leq x \leq c \qquad (1.17)$$

for as large a value of k as possible. Then we define the *Padé approximation* of order (n, m), or the (n, m) Padé approximation to $f(x)$, which we may consider to be the rational function analogue of the Taylor polynomial. If $f(x) \in C[-c, c]$, then such an approximation exists, since we can take $k \geq 0$. If $f(x) \in C^{(n+1)}$ $[-c, c]$ then we must have $k \geq n+1$ for we can take $P_n(x)$ to be the Taylor polynomial, and $Q_m(x) = 1$. Usually, if $f(x) \in C^{m+n+1}[-c, c]$ we can take $k = m + n + 1$, the number of degrees of freedom in $R(x)$ (but see exercise 33).

Example We will find the $(1, 1)$ Padé approximation to e^x.

$$\left| e^x - \frac{a_0 + a_1 x}{b_0 + b_1 x} \right| = \left| 1 + x + \frac{x^2}{2!} + \frac{x^3}{3!} + \ldots - \frac{a_0 + a_1 x}{b_0 + b_1 x} \right|$$

$$= (b_0 + b_1 x)^{-1} \left| \left(1 + x + \frac{x^2}{2!} + \ldots \right)(b_0 + b_1 x) - (a_0 + a_1 x) \right|$$

since $b_0 + b_1 x$ cannot be zero on $[-c, c]$ and we may assume that it is positive,

$$= (b_0 + b_1 x)^{-1} |b_0 + b_1 x + b_0 x + b_1 x^2 + \ldots - a_0 - a_1 x|.$$

Now we can eliminate the coefficient of x^i, $i = 0, 1, 2, \ldots$ in the expression between the modulus signs by setting

$$b_0 - a_0 = 0$$
$$b_1 + b_0 - a_1 = 0$$
$$b_1 + \frac{b_0}{2} = 0$$
$$\vdots$$

Normalizing so that $b_0 = 1$ (it cannot be zero) gives

$$a_0 = 1$$
$$a_1 = \tfrac{1}{2}$$
$$b_1 = -\tfrac{1}{2}$$

and so we can write

$$\left| e^x - \frac{1 + \tfrac{1}{2}x}{1 - \tfrac{1}{2}x} \right| \leq M|x^3|,$$

and we have obtained the required Padé approximation.

This example shows that if $f(x)$ can be expanded in a Taylor series about $x = 0$, then the coefficients of the (n, m) Padé approximation may be obtained by solving an appropriate system of equations. This can be made precise.

Theorem 1.11   Let $f(x) \in C^{n+m+1}[-c, c]$. Then if (1.17) cannot be satisfied with $k > n + m + 1$, the coefficients of the (n, m) Padé approximation to $f(x)$ satisfy the homogeneous equations

$$\sum_{i=0}^{j} \frac{D^i f(0)}{i!} b_{j-i} = a_j \qquad j = 0, 1, \dots, k-1,$$

where $a_{n+p} = b_{m+p} = 0$ if $p \geq 1$.

Proof   We may assume that the denominator $Q_m(x) > 0$ in $[-c, c]$, and so

$$\left| f(x) - \frac{P_n(x)}{Q_m(x)} \right| = Q_m(x)^{-1} |f(x) Q_m(x) - P_n(x)|.$$

By the assumption on $f(x)$, we may write

$$f(x) = P_{m+n}{}^*(x) + E(x)$$

where $P_{m+n}{}^*(x)$ is the Taylor polynomial of degree $m + n$ about $x = 0$ and

$$\| E(x) \|_\infty \leq \frac{1}{(m+n+1)!} |x|^{m+n+1} \| D^{m+n+1} f \|_\infty,$$

using (1.16). Thus

$$\left| f(x) - \frac{P_n(x)}{Q_m(x)} \right| \leq K_1 |P_{m+n}{}^*(x) Q_m(x) - P_n(x)| + K_2 |x|^{m+n+1}$$

for some constants K_1 and K_2, and so the required inequality is satisfied by choosing $P_n(x)$ and $Q_m(x)$ so that

$$P_{m+n}{}^*(x) Q_m(x) - P_n(x) = O(x^{m+n+1}),$$

where $A = O(h)$ means that $\lim_{h \to 0} A/h$ is finite. Expressing each term on the left hand side as an infinite series (with appropriate zero coefficients) and using the Cauchy rule for multiplication, we obtain the result on equating coefficients to zero. The details are left as an exercise. $\square$

Exercises

28.   Define $F(x)$ by

$$F(x) = f(x) - P_n(x) - (f(z) - P_n(z)) \frac{w(x)}{w(z)},$$

where $w(x) = \prod_{j=1}^{n+1} (x - x_j)$. Use Rolle's theorem on $D^n F$ (the nth derivative of F) to obtain (1.14).

29.   Use an argument similar to that of the previous exercise to obtain (1.15). Generalize this result, so that, for example, (1.16) is included.

30. Prove that

(i) $\displaystyle\sum_{i=1}^{n+1} l_i(x) = 1$

(ii) $l_i(x) = \dfrac{w(x)}{(x - x_i)Dw(x_i)}$ where $w(x)$ is as in exercise 28.

(iii) $\displaystyle\sum_{i=1}^{n+1} h_i(x) = 1$

(iv) $\displaystyle\sum_{i=1}^{n+1} [x_i h_i(x) + h_i' (x)] = x$

(v) $\displaystyle\sum_{i=1}^{n+1} (x - x_i) l_i^2 (x) Dl_i(x_i) = 0$

31. Determine the inverse of the matrix in Theorem 1.9, by making use of part (ii) of the previous exercise.

32. Determine the (2, 2) Padé approximation to $\cos(x)$.

33. It is possible for k in (1.17) to be less than $m + n + 1$. Illustrate this by finding the (1, 1) Padé approximation to the Bessel function

$$J_0(2x) = \sum_{i=0}^{\infty} (-1)^i \left(\frac{x^i}{i!}\right)^2 .$$

34. Consider the application of the trapezoidal rule

$$y_{n+1} = y_n + \frac{h}{2}[f(x_n, y_n) + f(x_{n+1}, y_{n+1})]$$

to the ordinary differential equation

$$Dy = f(x, y) \qquad y(0) = 1,$$

in the case where $f(x, y) = \lambda y, \lambda$ a constant, where h is the step length. Show that y_{n+1}/y_n gives the (1, 1) Padé approximation to $y(x_{n+1})/y(x_n)$. Consider also the effect of applying the backward Euler method: $y_{n+1} = y_n + hf(x_{n+1}, y_{n+1})$.

35. The Taylor polynomial of degree n about $x = 0$ interpolates to the function and its derivatives to order n. What can you say about the interpolating properties of Padé approximants?

2

The Chebyshev solution of an overdetermined system of linear equations

2.1 Introduction

In this chapter, we consider approximation in the space R^m normed with the L_∞ norm, or more popularly, Chebyshev norm (also referred to as the minimax, maximum, or uniform norm)

$$\|\mathbf{r}\| = \max_{1 \leqslant i \leqslant m} |r_i|.$$

Approximation with respect to this norm seems first to have been proposed by Laplace in 1799 in a study of inconsistent linear equations. The association with the name of the Russian mathematician P. L. Chebyshev (this is the popularly accepted English spelling, though many alternatives exist) is a consequence of his systematic investigation into a class of problems of this type, which was initiated in the 1850s. The basic problem to be examined here can be posed as follows:

$$\text{find } \mathbf{a} \in R^n \text{ to minimize } \quad \|\mathbf{r}(\mathbf{a})\| \tag{2.1}$$

where

$$\mathbf{r}(\mathbf{a}) = \mathbf{b} - A\mathbf{a}, \tag{2.2}$$

with A a given $m \times n$ matrix, and $\mathbf{b}$ a given vector in R^m. The explicit dependence of $\mathbf{r}$ (and other quantities) on $\mathbf{a}$ will often be suppressed, when no confusion can arise. In the abstract formulation of Chapter 1, $S = R^m$, M is the subspace of vectors of the form $A\mathbf{a}$, $\mathbf{a} \in R^n$, and g is the vector $\mathbf{b}$. The existence of a solution to (2.1) is guaranteed, by the previous analysis, although the uniqueness question is not trivial, as the space is not strictly convex. We begin by considering a simple example.

Example $m = 3$, $n = 1$,

$$A = \begin{bmatrix} 2 \\ \frac{1}{2} \\ -1 \end{bmatrix}, \qquad b = \begin{bmatrix} 2 \\ \frac{3}{2} \\ 0 \end{bmatrix}.$$

$$\| \mathbf{r} \| = \max \{ |2 - 2a_1|, |\tfrac{3}{2} - \tfrac{1}{2}a_1|, |a_1| \}.$$

In Figure 4 the individual functions appearing here are graphed, and it is clear from this that the values of $\| \mathbf{r} \|$ as a function of a_1 are taken on AB, BC, CD, DE. The minimum value is therefore at the point C, which is a vertex of the convex region mapped out by $\| \mathbf{r} \|$ defined by the second and third components of $\mathbf{r}$. At this point, $a_1 = 1$ and $\| \mathbf{r} \| = 1$, with $\| \mathbf{r} \| = r_2 = r_3$, $\| \mathbf{r} \| > r_1$.

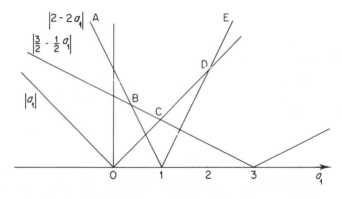

Figure 4

This simple example illustrates an elementary way of solving (2.1) for the case $n = 1$. More importantly, however, it suggests that for general n, significance is attached to vertices of the convex region in R^{n+1} defined by $\| \mathbf{r} \|$, such vertices being points at which $(n + 1)$ components of $\mathbf{r}$ have the same absolute value. It also shows that non-uniqueness of the solution is associated with the convex region defined by $\| \mathbf{r} \|$ having a base, which, geometrically speaking, is horizontal. We will place these ideas on a more solid footing, and begin by considering in the next section a precise characterization of solutions to (2.1).

Exercise

1. Obtain graphical solutions to the following problems with $n = 1$, $m = 3$.

(i) $A = \begin{bmatrix} 1 \\ 0 \\ -1 \end{bmatrix}$ $\quad \mathbf{b} = \begin{bmatrix} 1 \\ 2 \\ 1 \end{bmatrix}$

(ii) $A = \begin{bmatrix} 1 \\ 0 \\ -1 \end{bmatrix}$ $\quad \mathbf{b} = \begin{bmatrix} 1 \\ 1 \\ 1 \end{bmatrix}$

(iii) $A = \begin{bmatrix} 1 \\ 2 \\ -1 \end{bmatrix}$ $\quad \mathbf{b} = \begin{bmatrix} 3 \\ 1 \\ 0 \end{bmatrix}.$

2.2 Characterization of solutions

It is convenient to use $\overline{I}(= \overline{I}(\mathbf{a}))$ to denote the set of indices i corresponding to components of $\mathbf{r}$ which (in modulus) equal $\| \mathbf{r} \|$. We will also use θ_i to denote sign(r_i), where we define sign(α) to be $+1$, 0, -1 when $\alpha > 0$, $= 0$, < 0 respectively. Thus

$$r_i = \theta_i \| \mathbf{r} \| \qquad i \in \overline{I}.$$

Theorem 2.1 A vector $\mathbf{a} \in R^n$ solves (2.1) if and only if there exists $I \subset \overline{I}$ containing at most $(n+1)$ indices, and a nontrivial vector $\lambda \in R^m$ such that

$$\lambda_i = 0 \qquad i \notin I$$
$$A^T \lambda = \mathbf{0}$$
$$\lambda_i \theta_i \geqslant 0 \qquad i \in I.$$

(If $\| \mathbf{r} \| = 0$, this result is trivial, so assume not).

Proof 1 In the notation of Theorem 1.7, we have

$$V(\mathbf{r}) = \mathrm{conv}\{ \theta_i \mathbf{e}_i, \ i \in \overline{I} \}.$$

Thus $\mathbf{v} \in V(\mathbf{r})$ implies that

$$\mathbf{v} = \sum_{i \in \overline{I}} \mu_i \theta_i \mathbf{e}_i,$$

where $\mu_i \geqslant 0$, $i \in \overline{I}$, $\sum_{i \in \overline{I}} \mu_i = 1$. The required result follows as an immediate consequence of Theorem 1.7 and Theorem 1.4.

Proof 2 Suppose that $\mathbf{a}$ does not solve (2.1), but the conditions of the theorem are satisfied. Then there exists $\mathbf{c} \in R^n$ such that

$$\| \mathbf{r}(\mathbf{a} + \mathbf{c}) \| < \| \mathbf{r}(\mathbf{a}) \|.$$

In particular

$$|r_i(\mathbf{a}+\mathbf{c})| < |r_i(\mathbf{a})| \qquad i \in I,$$

or

$$|r_i(\mathbf{a}) - \boldsymbol{\alpha}_i{}^T \mathbf{c}| < |r_i(\mathbf{a})| \qquad i \in I,$$

where $\boldsymbol{\alpha}_i{}^T$ denotes the ith row of A. Thus

$$\theta_i \boldsymbol{\alpha}_i{}^T \mathbf{c} > 0 \qquad i \in I. \tag{2.3}$$

Now the conditions of the theorem can be written

$$\sum_{i \in I} \mu_i \theta_i \boldsymbol{\alpha}_i{}^T \mathbf{c} = 0$$

where $\mu_i = \lambda_i \theta_i \geqslant 0$, $i \in I$. In view of (2.3) and the nontriviality of λ, this gives a contradiction.

Now let $\mathbf{a}$ be a solution to (2.1), and assume that the conditions of the theorem are not satisfied. Then

$$\mathbf{0} \notin D$$

where D is the convex hull of vectors $\theta_i \boldsymbol{\alpha}_i$, $i \in \overline{I}$ and so by Theorem 1.5, there exists $\mathbf{c} \in R^n$ such that

$$\theta_i \boldsymbol{\alpha}_i{}^T \mathbf{c} > 0 \qquad i \in \overline{I}.$$

Thus for $i \in \overline{I}$,

$$\begin{aligned}
|r_i(\mathbf{a}+\gamma\mathbf{c})| &= |\theta_i\|\mathbf{r}(\mathbf{a})\| - \gamma\boldsymbol{\alpha}_i{}^T\mathbf{c}| \\
&= |\,\|\mathbf{r}(\mathbf{a})\| - \gamma\theta_i\boldsymbol{\alpha}_i{}^T\mathbf{c}| \\
&< \|\mathbf{r}(\mathbf{a})\|
\end{aligned}$$

provided that

$$0 < \gamma < \frac{2\|\mathbf{r}(\mathbf{a})\|}{\max_{i \in \overline{I}}(\theta_i\boldsymbol{\alpha}_i{}^T\mathbf{c})}.$$

For $i \notin \overline{I}$, $\gamma > 0$,

$$\begin{aligned}
|r_i(\mathbf{a}+\gamma\mathbf{c})| &= |r_i(\mathbf{a}) - \gamma\boldsymbol{\alpha}_i{}^T\mathbf{c}| \\
&\leqslant \max_{i \notin \overline{I}}|r_i(\mathbf{a})| + \gamma \max_{i \notin \overline{I}} |\boldsymbol{\alpha}_i{}^T\mathbf{c}| \\
&< \|\mathbf{r}(\mathbf{a})\|
\end{aligned}$$

provided that

$$0 < \gamma < (\|\mathbf{r}(\mathbf{a})\| - \max_{i \notin \overline{I}} |r_i(\mathbf{a})|)/\max_{i \notin \overline{I}} |\boldsymbol{\alpha}_i{}^T\mathbf{c}|.$$

Thus if $\gamma > 0$ is small enough, the fact that $\mathbf{a}$ solves (2.1) is contradicted, and the result is proved. $\qquad\square$

Corollary 1 Let **a** solve (2.1). Then **a** solves an L_∞ problem in R^{n+1} obtained by restricting the components of **r** to a particular $n+1$. If A has rank t, then the components of **r** may be restricted to a particular $t+1$.

Proof If A has rank t, we can always choose I to contain $s \leqslant t+1$ indices (show this). Thus, by the theorem, **a** solves the L_∞ problem in R^{n+1} (or R^{t+1}) obtained by restricting the components of **r** to any $n+1$ (or $t+1$) which include I. □

Corollary 2 Let **a** solve (2.1) and let I be such that $\lambda_i \neq 0$, $i \in I$. Then

$$r_i(\mathbf{d}) = r_i(\mathbf{a}) \qquad i \in I,$$

for all solutions **d** to (2.1).

Proof Let $h = \|\mathbf{r}(\mathbf{a})\|$, and let **d** be any other solution to (2.1). If $h = 0$ the result is trivial, so assume that $h > 0$. Then by the theorem,

$$\sum_{i \in I} \lambda_i \boldsymbol{\alpha}_i = \mathbf{0} \qquad \lambda_i \theta_i > 0, \, i \in I,$$

and so

$$h \sum_{i \in I} |\lambda_i| = \left| \sum_{i \in I} \lambda_i r_i(\mathbf{a}) \right|$$

$$= \left| \sum_{i \in I} \lambda_i b_i \right|$$

$$= \left| \sum_{i \in I} \lambda_i r_i(\mathbf{d}) \right|$$

$$\leqslant \sum_{i \in I} |\lambda_i| \, |r_i(\mathbf{d})|$$

$$\leqslant h \sum_{i \in I} |\lambda_i|.$$

Thus equality holds throughout, and the result follows. □

Theorem 2.2 If A has rank t, a solution **a** to (2.1) always exists with $\overline{I}$ containing at least $(t+1)$ indices.

Proof Let A have rank t, and let **a** solve (2.1) with $\overline{I}(\mathbf{a})$ containing $s < t+1$ indices. Since the corresponding rows of A are linearly dependent (by Theorem 2.1) there exists a nontrivial vector $\mathbf{c} \in R^n$ with

$$\boldsymbol{\alpha}_i^T \mathbf{c} = 0 \qquad i \in \overline{I}(\mathbf{a}). \tag{2.4}$$

Thus

$$|r_i(\mathbf{a} + \gamma \mathbf{c})| = |r_i(\mathbf{a})| = \|\mathbf{r}(\mathbf{a})\| \qquad i \in \overline{I}(\mathbf{a})$$
$$|r_i(\mathbf{a} + \gamma \mathbf{c})| = |r_i(\mathbf{a}) - \gamma \boldsymbol{\alpha}_i^T \mathbf{c}| \qquad i \notin \overline{I}(\mathbf{a}).$$

30

Further, by the rank condition on A, for at least one $\mathbf{c}$ satisfying (2.4) we must have

$$\boldsymbol{\alpha}_j^T \mathbf{c} = \delta \neq 0 \qquad \text{for some } j \notin \overline{I}(\mathbf{a}).$$

Thus, we can increase $|\gamma|$ away from zero until the first index not in $\overline{I}(\mathbf{a})$ is such that

$$|r_i(\mathbf{a}+\gamma\mathbf{c})| = \|\mathbf{r}(\mathbf{a}+\gamma\mathbf{c})\| = \|\mathbf{r}(\mathbf{a})\|,$$

and so $\mathbf{a}+\gamma\mathbf{c}$ is a solution to (2.1) with $\overline{I}(\mathbf{a}+\gamma\mathbf{c})$ containing at least one more index than $\overline{I}(\mathbf{a})$. We may continue this process until no suitable $\mathbf{c}$ can be obtained, and the result follows. □

Corollary The submatrix of A consisting of the rows of A corresponding to $\overline{I}(\mathbf{a})$ must have rank t for some solution $\mathbf{a}$ to (2.1).

These results show that subsets of the indices $\{1, 2, \ldots, m\}$ containing just $(t+1)$ indices are of importance. A particular subset associated through Theorem 2.1 with a solution to (2.1) (an *extremal* subset) may be distinguished from all other such subsets by the following result, due essentially to de la Vallée Poussin (1911).

Theorem 2.3 Let A have rank t, and let J be any subset of $(t+1)$ indices from $\{1, 2, \ldots, m\}$. Then

$$\min_{\mathbf{a}} \{\max_{i \in J} |r_i(\mathbf{a})|\} \leq \min_{\mathbf{a}} \|\mathbf{r}(\mathbf{a})\|, \tag{2.5}$$

with equality holding when J is an extremal subset.

Proof Let $\min_{\mathbf{a}} \|\mathbf{r}(\mathbf{a})\| = h$, and let $\mathbf{d}$ be any solution to (2.1). Then

$$|r_i(\mathbf{d})| \leq h \qquad i = 1, 2, \ldots, m,$$

so that

$$\max_{i \in J} |r_i(\mathbf{d})| \leq h.$$

The result then follows using Corollary 1 of Theorem 2.1. □

This theorem is important in that it motivates a class of numerical methods for (2.1), based on solving a sequence of problems in R^{t+1} to find an extremal subset. From a practical point of view, an immediate difficulty is that in general no information is available *a priori* on the rank of A, so that t is not known in advance. It is necessary, therefore, to either require an assumption that A has rank n, or else to establish the rank of A as the first stage in the calculation, so that $(n-t)$ columns of A can be effectively deleted. The basic subproblem, then, is (without loss of generality) the solution of a problem in R^{n+1} with a matrix A of rank n, and this is considered in Section 2.4. Firstly, however, we turn to the problem of uniqueness of the solution to (2.1).

Exercises

2. Verify the results of the last section for the solutions to the problems in exercise 1.

3. Let

$$A = \begin{bmatrix} 1 & 2 \\ 0 & 1 \\ 1 & 2 \end{bmatrix}, \quad b = \begin{bmatrix} 1 \\ 2 \\ 0 \end{bmatrix}.$$

Determine all the vertices of $\|\mathbf{r}\|$ and hence solve the L_∞ problem. Repeat for

$$A = \begin{bmatrix} 1 & 1 \\ 2 & 2 \\ 3 & 3 \\ 4 & 4 \end{bmatrix}, \quad b = \begin{bmatrix} 1 \\ 1 \\ 1 \\ 1 \end{bmatrix}$$

and for both problems verify the results of the last section.

4. Determine the best Chebyshev approximation to x^2 on the 3 points 0, 1, and 2 by a constant.

5. Let $\mathbf{d} \in R^n$, and let J be a set of indices from $\{1, 2, \ldots, m\}$ such that

$$\sum_{i \in J} \lambda_i \boldsymbol{\alpha}_i = \mathbf{0},$$
$$\lambda_i \text{ sign } (r_i(\mathbf{d})) \geq 0 \qquad i \in J,$$

for some numbers λ_i, $i \in J$, not all zero. Prove that

$$\min_{i \in J} |r_i(\mathbf{d})| \leq \min_{\mathbf{a}} \|\mathbf{r}(\mathbf{a})\|.$$

2.3 Uniqueness of the solution

Corollary 2 to Theorem 2.1 points the way to the kind of condition on A required to guarantee uniqueness of the solution to (2.1): for an appropriate subset I, we must ensure that

$$r_i(\mathbf{a}) = r_i(\mathbf{d}) \qquad i \in I,$$

or equivalently

$$\boldsymbol{\alpha}_i^T(\mathbf{a} - \mathbf{d}) = 0 \qquad i \in I,$$

has the implication $\mathbf{a} = \mathbf{d}$. This is achieved by making the rows of A satisfy a strong condition of linear independence known as the Haar condition.

Definition 2.1 An $m \times n$ matrix A, where $m \geqslant n$, satisfies the *Haar condition* if every $n \times n$ submatrix is nonsingular.

Theorem 2.4 If A satisfies the Haar condition, the solution to (2.1) is unique.

Proof Let **a** be a solution to (2.1). Then, by Theorem 2.1, there exists $I \subset \bar{I}(\mathbf{a})$, $\lambda \in R^m$ such that

$$\sum_{i \in I} \lambda_i \alpha_i = \mathbf{0}.$$

If A satisfies the Haar condition, I must contain exactly $(n+1)$ indices (exercise 6), and so if **d** is also a solution, Corollary 2 to Theorem 2.1 gives that

$$\alpha_i{}^T(\mathbf{a} - \mathbf{d}) = 0 \qquad i \in I.$$

The Haar condition is therefore contradicted unless $\mathbf{a} = \mathbf{d}$, and so the result is proved.   □

In fact some information is available about how *quickly* $\|\mathbf{r}(\mathbf{a})\|$ increases as **a** moves away from the (unique) solution to (2.1). We require the following definition.

Definition 2.2 The solution $\mathbf{a} \in R^n$ to (2.1) is *strongly unique* if there exists $\gamma > 0$ such that

$$\|\mathbf{r}(\mathbf{d})\| \geqslant \|\mathbf{r}(\mathbf{a})\| + \gamma \|\mathbf{d} - \mathbf{a}\|_2$$

for all $\mathbf{d} \in R^n$, where $\|.\|_2$ is the L_2 norm on R^n.

The following theorem was essentially first given by Newman and Shapiro (1962).

Theorem 2.5 Let A satisfy the Haar condition. Then the solution to (2.1) is strongly unique.

Proof Let A satisfy the Haar condition. Then we may define

$$\min_{\|\mathbf{d}\|_2 = 1} \|A\mathbf{d}\| = \delta > 0$$

since the minimum of a continuous function on a compact set is attained. Let **a** solve (2.1). Then if $\|\mathbf{r}(\mathbf{a})\| = 0$ we have

$$\|\mathbf{r}(\mathbf{d})\| = \|\mathbf{b} - A\mathbf{d}\|$$
$$= \|A(\mathbf{a} - \mathbf{d})\|,$$

so that we can satisfy the required inequality with $\gamma = \delta$. Assume that $\|\mathbf{r}(\mathbf{a})\| > 0$. By Theorem 2.2, there are $(n+1)$ indices $I \subset \bar{I}(\mathbf{a})$ and $\lambda \in R^m$ so that

$$\sum_{i \in I} \lambda_i \alpha_i = \mathbf{0},$$

with $\lambda_i \theta_i > 0$, $i \in I$, where $\theta_i = \text{sign} (r_i(\mathbf{a}))$, $i \in I$. Let $\mathbf{c} \in R^n$ be such that $\|\mathbf{c}\|_2 = 1$. Now

$$\sum_{i \in I} \lambda_i \boldsymbol{\alpha}_i{}^T \mathbf{c} = 0$$

with $\boldsymbol{\alpha}_i{}^T \mathbf{c}$ nonzero for at least 2 values of $i \in I$. It follows that

$$\max_{i \in I} \theta_i \boldsymbol{\alpha}_i{}^T \mathbf{c} > 0$$

and so we may define a positive number γ by

$$\gamma = \min_{\|\mathbf{c}\|_2 = 1} \max_{i \in I} \theta_i \boldsymbol{\alpha}_i{}^T \mathbf{c},$$

since we are minimizing a continuous function on a compact set. Let $\mathbf{d} \in R^n$ be arbitrary. Then if $\mathbf{d} = \mathbf{a}$, the required inequality follows immediately. Otherwise let

$$\mathbf{c} = \frac{\mathbf{a} - \mathbf{d}}{\|\mathbf{a} - \mathbf{d}\|_2}$$

from which it follows that there is an index $k \in I$ such that

$$\theta_k \boldsymbol{\alpha}_k{}^T \mathbf{c} \geqslant \gamma.$$

Thus

$$\begin{aligned}
\|\mathbf{r}(\mathbf{d})\| &\geqslant \theta_k r_k(\mathbf{d}) \\
&= \theta_k r_k(\mathbf{a}) + \theta_k \boldsymbol{\alpha}_k{}^T(\mathbf{a} - \mathbf{d}) \\
&= \theta_k r_k(\mathbf{a}) + \theta_k \boldsymbol{\alpha}_k{}^T \mathbf{c} \|\mathbf{d} - \mathbf{a}\|_2 \\
&\geqslant \|\mathbf{r}(\mathbf{a})\| + \gamma \|\mathbf{d} - \mathbf{a}\|_2.
\end{aligned}$$

$\square$

The Haar condition is not necessary for strong uniqueness, as the following example shows.

Example

$$A = \begin{bmatrix} 1 \\ 0 \end{bmatrix}, \qquad \mathbf{b} = \begin{bmatrix} 0 \\ 0 \end{bmatrix}.$$

The solution is $a_1 = 0$, and

$$\|\mathbf{r}(\mathbf{d})\| \geqslant \|\mathbf{d}\|_2 \qquad \text{for all } \mathbf{d} \in R^1.$$

In fact, uniqueness of the solution to (2.1) is all that is required to guarantee strong uniqueness, as may be shown by appropriate modification to Theorem 2.5 (exercise 7). The property of strong uniqueness is a consequence of the solution being attained at a single vertex of $\|\mathbf{r}(\mathbf{a})\|$ and corresponds geometrically to the existence of a cone in R^{n+1}, with vertex at

$$\begin{bmatrix} \mathbf{a}^* \\ \|\mathbf{r}(\mathbf{a}^*)\| \end{bmatrix}$$

where $\mathbf{a}^*$ is the solution to (2.1), whose surface separates the horizontal plane through this point and the convex region defined by $\|\mathbf{r}(\mathbf{a})\|$ and has contact only at the common point. This situation may be contrasted with the problem of minimizing a smooth (differentiable) function, for example a strictly convex L_p norm, when strong uniqueness is not generally possible. The two cases are illustrated in Figure 5.

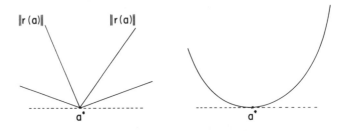

Figure 5

Exercises

6. Let A be an $(n+1) \times n$ matrix which satisfies the Haar condition. Show that there exists a vector $\lambda \in R^{n+1}$, unique up to a scalar multiplier, such that

$$A^T \lambda = 0,$$

with $\lambda_i \neq 0$, $i = 1, 2, \ldots, n+1$.

7. Let (2.1) have a unique solution. Prove that the solution is strongly unique. (In Theorem 2.5, show that if

$$\gamma = \min_{\|\mathbf{c}\|_2 = 1} \max_{i \in \bar{I}(\mathbf{a})} \theta_i \alpha_i^T \mathbf{c},$$

then $\gamma \leqslant 0$ leads to a contradiction of uniqueness.)

2.4 Solutions to subproblems in R^{n+1}

One way of obtaining the minimum value of $\|\mathbf{r}\|$ is to identity an extremal subset, and solve a Chebyshev approximation problem on that subset. Such a set may be obtained by computing solutions on all possible such subsets. More systematic methods, however, are available, and we consider this later; meantime we examine how solutions to problems in R^{n+1} may be obtained efficiently, and the relationship of solutions on extremal subsets to solutions of (2.1). Suppose we have

$$r_i = b_i - \boldsymbol{\alpha}_i^T \mathbf{a} \qquad i \in J = \{1, 2, \ldots, n+1\}, \tag{2.6}$$

where $\max_{i \in J} |r_i|$ is to be minimized. Now any solution will be such that this maximum is attained at a set of indices $I \subset J$. Thus consider the system of equations

$$b_i - \boldsymbol{\alpha}_i^T \mathbf{a} = \hat{\theta}_i h \qquad i \in J \tag{2.7}$$

where $\hat{\theta}_i$ are numbers such that $|\hat{\theta}_i| \leqslant 1, i \in J$. Then clearly $\max_{i \in J} |r_i| = |h|$ if $|\hat{\theta}_i| = 1$ for some $i \in J$. Let $\boldsymbol{\lambda} \in R^{n+1}$ be such that $\boldsymbol{\lambda} \neq \mathbf{0}$ with

$$\sum_{i \in J} \lambda_i \boldsymbol{\alpha}_i = \mathbf{0}, \ \boldsymbol{\lambda}^T \mathbf{b} \geqslant 0. \tag{2.8}$$

Then

$$\boldsymbol{\lambda}^T \mathbf{b} = h \boldsymbol{\lambda}^T \hat{\boldsymbol{\theta}},$$

and so

$$|h| = \boldsymbol{\lambda}^T \mathbf{b} / |\boldsymbol{\lambda}^T \hat{\boldsymbol{\theta}}|.$$

The right hand side is minimized by choosing $\hat{\boldsymbol{\theta}}$ to be such that $\hat{\theta}_i = \text{sign}(\lambda_i)$, $\lambda_i \neq 0$. Clearly this choice also corresponds to the satisfaction of the conditions of Theorem 2.1 in this case, since $\theta_i \equiv \hat{\theta}_i, \lambda_i \neq 0$. To summarize, therefore, we may obtain a solution to the problem defined by (2.6) by the following procedure:

(i) find a nontrivial $\boldsymbol{\lambda} \in R^{n+1}$ satisfying (2.8),
(ii) set $\hat{\theta}_i = \text{sign}(\lambda_i), \lambda_i \neq 0$,

$$|\hat{\theta}_i| \leqslant 1, \qquad \text{but otherwise arbitrary if } \lambda_i = 0,$$

(iii) solve (2.7) for $\mathbf{a}$ and h.

There is, of course, the assumption made here that the $(n+1) \times (n+1)$ matrix defined by

$$\begin{bmatrix} \boldsymbol{\alpha}_1^T & \hat{\theta}_1 \\ \vdots & \vdots \\ \vdots & \vdots \\ \boldsymbol{\alpha}_{n+1}^T & \hat{\theta}_{n+1} \end{bmatrix} \tag{2.9}$$

is nonsingular, so the system of equations (2.7) has a solution. If the $(n+1) \times n$ matrix formed by the first n columns of the matrix (2.9) has rank n, then the nonsingularity may be demonstrated: the details are left as exercise 8.

If J is an extremal subset of the original problem in R^m, then h will give the minimum value of $\|\mathbf{r}\|$. For any solution to the problem on this extremal subset to be a solution to the original problem, we clearly require the satisfaction of the inequalities

$$|r_i| \leq h \qquad i = 1, 2, \ldots, m,$$

and this need not follow for all solutions.

Example $m = 4, n = 2,$

$$A = \begin{bmatrix} 1 & 2 \\ 2 & 3 \\ 2 & 4 \\ 1 & 0 \end{bmatrix}, \qquad b = \begin{bmatrix} 1 \\ 1 \\ 1 \\ 1 \end{bmatrix}.$$

An extremal subset is given by $J = \{1, 2, 3\}$ and here we can take $\lambda_1 = 2, \lambda_2 = 0,$ $\lambda_3 = -1$. Thus $\hat{\theta}_1 = +1, \hat{\theta}_3 = -1$. The system of equations (2.7) is thus

$$\begin{aligned} 1 - a_1 - 2a_2 &= h \\ 1 - 2a_1 - 3a_2 &= \hat{\theta}_2 h \\ 1 - 2a_1 - 4a_2 &= -h \end{aligned}$$

giving the solution $a_1 = -2\hat{\theta}_2/3$, $a_2 = (1 + \hat{\theta}_2)/3$, $h = \frac{1}{3}$, which, for any $\hat{\theta}_2$ satisfying $|\hat{\theta}_2| \leq 1$ solves the problem on this subset. Now

$$|r_4| = |1 + 2\hat{\theta}_2/3| \leq \frac{1}{3}$$

provided that $\hat{\theta}_2 = -1$. Thus the (unique) solution to the original problem is given by $a_1 = 2/3, a_2 = 0.$

Exercises

8.   Prove that the matrix (2, 9) is nonsingular if the first n columns are linearly independent (have rank n). (Assume singularity, and obtain a contradiction.)

9.   If it is only assumed that the matrix whose rows are $\alpha_i^T, i \in J$ has rank $t \leq n$, what are the consequences for the analysis of this section?

10.   When $m = n + 1$, show that the Haar condition on A is necessary as well as sufficient for the solution of (2.1) to be unique, provided that $\| r(a) \| > 0.$

11.   Solve the problem (2.1) with

$$A = \begin{bmatrix} 1 & -1 \\ 2 & 3 \\ 3 & 1 \end{bmatrix}, \qquad b = \begin{bmatrix} 7 \\ 5 \\ 1 \end{bmatrix},$$

by the method of this section. Also solve the problem of exercise 3.

12. Solve the problem (2.1) with

$$A = \begin{bmatrix} 1 & 2 \\ 2 & 4 \\ -1 & -2 \end{bmatrix}, \quad b = \begin{bmatrix} 1 \\ 2 \\ 1 \end{bmatrix}$$

(see exercise 9).

13. Consider the problem of approximating to x^2 by a linear combination of x and e^x on the 3 points 0, 1, and 2, using the Chebyshev norm. Find the sign pattern of the errors at these points, and write down and solve the system of equations defining the solution.

14. Experimentally determined values of a function f are given in the table for given values of x. Assuming that there exists a relation of the form $f = ax$, find the best value of a in the Chebyshev sense.

x	1	2	3	4
f	1.1	1.6	2.1	2.5

15. Illustrate graphically the situation in exercise 13, and hence confirm the solution.

2.5 Ascent methods

The results of the previous sections, in particular Theorem 2.3, suggest that the identification of an extremal subset may be obtained by starting from any particular subset and changing to another in such a way that the new solution gives a larger value of $\max_{i \in J} |r_i|$. Since there are only a finite number of such subsets, ultimately we must reach a stage when no further increase is possible, and thus an extremal subset has been obtained. This is the basis of methods for solving (2.1) known as *ascent methods*. The earliest was due to Stiefel (1959) called the exchange method, in which a single index was exchanged at each iteration. In the original algorithm, it was assumed that A satisfied the Haar condition: each subproblem therefore has a unique solution, and the solution on the extremal subset is the solution to the original problem. The basis of the exchange method is given in the following theorem, the proof of which gives the essential details of its implementation.

Theorem 2.6 Let A satisfy the Haar condition, and let J be a subset of $\{1, 2, \ldots, m\}$ containing $(n + 1)$ indices. Then unless J is extremal, it is possible to exchange one index of J to form a new set J^* such that

$$h = \min_{a} \max_{i \in J} |r_i(a)| < \min_{a} \max_{i \in J^*} |r_i(a)| = h^*.$$

Proof Assume (without loss of generality) that $J = \{1, 2, \ldots, n+1\}$, that c is the solution on this subset, and that

$$|r_{n+2}(\mathbf{c})| = \max_{n+2 \leqslant i \leqslant m} |r_i(\mathbf{c})| > h. \qquad (2.10)$$

Let M be the $(n+2) \times n$ matrix formed by the first $n+2$ rows of A. Then M has rank n, and there are 2 linearly independent vectors $\mathbf{v}$ such that

$$\mathbf{v}^T M = \mathbf{0}. \qquad (2.11)$$

One such vector is $(\boldsymbol{\lambda}^T, 0)^T$, where $\boldsymbol{\lambda} \in R^{n+1}$ with

$$h = \sum_{i=1}^{n+1} \lambda_i b_i / \sum_{i=1}^{n+1} \lambda_i \theta_i,$$

and $\theta_i = \text{sign} \ (r_i(\mathbf{c}))$, $i = 1, 2, \ldots, n+1$. Now let

$$\mathbf{v} = \gamma \begin{bmatrix} \boldsymbol{\lambda} \\ 0 \end{bmatrix} + \begin{bmatrix} \boldsymbol{\mu} \\ \text{sign} \ (r_{n+2}(\mathbf{c})) \end{bmatrix}, \qquad (2.12)$$

where

$$\sum_{i=1}^{n+1} \mu_i \boldsymbol{\alpha}_i^T + \text{sign} \ (r_{n+2}(\mathbf{c})) \boldsymbol{\alpha}_{n+2}^T = \mathbf{0}.$$

Then $\mathbf{v}$ provides a second vector satisfying (2.11). Further

$$v_i r_i(\mathbf{c}) = \lambda_i r_i(\mathbf{c}) (\gamma + \mu_i / \lambda_i) \qquad i = 1, 2, \ldots, n+1$$

$$\geqslant 0$$

provided that $\gamma = \max_{1 \leqslant i \leqslant n+1} (-\mu_i / \lambda_i) = -\mu_j / \lambda_j$, say. With this choice of γ,

$$\sum_{\substack{i=1 \\ i \neq j}}^{n+2} v_i \boldsymbol{\alpha}_i = \mathbf{0},$$

with $v_i r_i(\mathbf{c}) > 0$, $i = 1, 2, \ldots, n+2$, $i \neq j$. Define $J^* = \{1, 2, \ldots, j-1, j+1, \ldots, n+2\}$. Then

$$h^* = \sum_{i \in J^*} v_i b_i / \sum_{i \in J^*} |v_i|$$

$$= \sum_{i \in J^*} v_i r_i(\mathbf{c}) / \sum_{i \in J^*} |v_i|$$

$$= \left[\sum_{\substack{i=1 \\ i \neq j}}^{n+1} |v_i| h + |v_{n+2}| |r_{n+2}(\mathbf{c})| \right] \bigg/ \sum_{i \in J^*} |v_i|$$

$$> h \qquad \text{by (2.10)}. \qquad \qquad \square$$

Example $m = 4, n = 2,$

$$A = \begin{bmatrix} 1 & 2 \\ 2 & 1 \\ 1 & 0 \\ -1 & 1 \end{bmatrix}, \qquad b = \begin{bmatrix} 1 \\ -1 \\ 1 \\ 2 \end{bmatrix}.$$

Let $J = \{1, 2, 3\}$. We can take $\lambda_1 = 1, \lambda_2 = -2, \lambda_3 = 3$, with $c_1 = c_2 = 0, h = 1$. Now $r_4 = 2$, so μ must satisfy

$$\mu_1 + 2\mu_2 + \mu_3 - 1 = 0$$
$$2\mu_1 + \mu_2 \qquad + 1 = 0.$$

We can take $\mu_1 = -1, \mu_2 = 1, \mu_3 = 0$, for example, which gives $j = 1$ and $J^* = \{2, 3, 4\}$. The completion of this example is left as an exercise.

The Haar condition is clearly necessary for the proof of the above theorem, to ensure that all the quotients μ_i / λ_i are finite. However, the Stiefel exchange method is in fact equivalent to a method based on a linear programming approach for which the weaker assumption that A has rank n is sufficient to guarantee that a solution to (2.1) is obtained by standard methods. To see the connection between (2.1) and a linear programming problem, let us set $z = \max\limits_{1 \leqslant i \leqslant m} |r_i|$. Then (2.1) is equivalent to

$$\text{minimize } z$$
$$\text{subject to } -z \leqslant r_i \leqslant z, i = 1, 2, \ldots, m,$$

or

$$\text{minimize } z$$
$$\text{subject to } \begin{bmatrix} A & e \\ -A & e \end{bmatrix} \begin{bmatrix} a \\ z \end{bmatrix} \geqslant \begin{bmatrix} b \\ -b \end{bmatrix},$$

where $e = (1, 1, \ldots, 1)^T$. This is a linear programming problem in the $(n+1)$ variables a, z. It is not, however, in a form suitable for the direct application of standard techniques, as the components of a are not constrained non-negative. Going to the dual formulation of the problem, however, gives

$$\begin{aligned} \text{maximize} \quad & (v - w)^T b \\ \text{subject to} \quad & A^T(v - w) = 0 \\ & e^T(v + w) \leqslant 1 \\ & v, w \geqslant 0. \end{aligned} \qquad (2.13)$$

This form is such that all variables are constrained non-negative. Additional advantages are that all but one of the constraints are equality (so that only one slack variable is required), and also the number of equations has been reduced to $(n+1)$ instead of $2m$. If A has rank n, then the matrix of the system of equalities which results on the insertion of a slack variable in (2.13) has rank $(n+1)$, and the standard simplex method can be applied directly.

One of the first to consider the (primal) linear programming formulation of the problem (2.1) was Zuhovitzski in the early 1950s (in Russian). The advantages to be gained in using the dual seem first to have been pointed out by Kelley (1958), who gives an application to curve fitting. Barrodale and Young (1966) give an algorithm using the standard simplex method and Bartels and Golub (1968) give a version which uses a numerically stable factorization procedure. The application of the simplex algorithm to the dual problem (2.13) in fact gives an exchange procedure which is precisely equivalent to the original Stiefel exchange method once only variables from $\mathbf{v}$ and $\mathbf{w}$ are present in the basis (see Stiefel, 1960; Osborne and Watson, 1967; where the precise details of the equivalence are given). At each step of the simplex method, there are $(n + 1)$ basic variables. The primal–dual relationship tells us that strict equality must therefore hold in the corresponding primal constraints. Thus (assuming $\mathbf{r} \neq \mathbf{0}$), $(n + 1)$ components of $\mathbf{r}$ are picked out and the current basic feasible solution just gives a solution on this particular subset. It is interesting that the well-known linear programming result that an optimal solution always exists at a vertex is just another way of stating the result of Theorem 2.2. Another connection which may be readily observed is the relationship between the vector of basic variables, and the nonzero components of the vector of the characterization theorem: each basic variable can be regarded as some $\lambda_j \theta_j$ with λ normalized so that $\sum |\lambda_i| = 1$. The details of this relation are left as an exercise.

If A satisfies the Haar condition, then this equivalence shows that the sequence of basic feasible solutions is nondegenerate, and so there is a strict increase in the value of the objective function at each step. If the Haar condition is not satisfied, the occurrence of degeneracy may cause cycling, i.e. the objective function may remain constant over a number of iterations so that an earlier basis is recovered. This problem can be dealt with by standard linear programming techniques; however, it is not regarded as being a likely occurrence, and usually no safety measures of this kind are included in computer programs.

Recent developments in the implementation of ascent methods have been in making the process both robust, numerically stable, and efficient. In particular, the algorithm of Barrodale and Phillips (1975a) may be applied to problems with an arbitrary matrix A, and converges to a solution in relatively few iterations. The method is a modification of the standard simplex method which takes advantage of the special structure of (2.13), in particular the ease with which a variable w_i may be interchanged with the corresponding variable v_i in the basis. This is a consequence of the following observation. Suppose that the basis matrix contains as $(n + 1)$st column $\mathbf{e}_{n+1}$ corresponding to the slack variable introduced in (2.13), and as rth column that corresponding to the variable v_k. Let the inverse of this matrix be denoted by

$$\begin{bmatrix} \mathbf{b}_1^T \\ \mathbf{b}_2^T \\ \vdots \\ \mathbf{b}_{n+1}^T \end{bmatrix}$$

so that $\mathbf{b}_i^T$ is the ith row. Then if w_k replaces v_k in the basis, the inverse basis matrix becomes

$$\begin{bmatrix} \mathbf{b}_1{}^T \\ \vdots \\ -\mathbf{b}_r{}^T \\ \vdots \\ \mathbf{b}_{n+1}^T + 2\mathbf{b}_r{}^T \end{bmatrix}$$

where only the rth and $(n+1)$st rows are altered (exercise 20). It is clear that w_k takes on the negative of the value previously taken by v_k.

Initially the n artificial variables and the slack variable are basic. Then in the first part of the algorithm, only variables v_i are allowed to become basic (in particular the variable v_k with coefficient in the objective function of largest *absolute* value) and only the artificial variables can become nonbasic (in particular that whose coefficient of v_k is largest in absolute value). The pivot may thus have the wrong sign, but since the right hand sides are zero, this is permissible. If A has rank t, then t such steps are performed. The $(t+1)$st step consists of again looking for the variable v_k to become basic in the same way. However, if the sign of the coefficient of this variable is the wrong sign for a maximization problem, then the corresponding variable w_k is actually chosen. The slack variable then becomes nonbasic; to ensure that this is possible (i.e. to retain feasibility) may require altering the signs of the coefficients of the variable to become basic in any or all of the first n rows, and this is achieved by interchanging some of the basic variables v_i for w_i first.

At this point, we have a basic feasible solution to (2.13) which corresponds to a point where $(t+1)$ primal constraints hold with equality. The second part of the method consists of iterating through further such 'vertices' by conventional simplex (or exchange) steps, which leave any artificial variables still in the basis unchanged, until optimality is achieved. A Fortran program for the method is given by Barrodale and Phillips (1975b), where more detailed information regarding the recommended implementation can be obtained. We illustrate by a simple example.

Example $m = 3$, $n = 2$,

$$A = \begin{bmatrix} 1 & 1 \\ 0 & 1 \\ 2 & 0 \end{bmatrix}, \qquad \mathbf{b} = \begin{bmatrix} 1 \\ 2 \\ 1 \end{bmatrix}.$$

To avoid the introduction of linear programming notation, for the purposes of the example we simply express the basic variables in terms of the nonbasic ones. It should be remembered throughout that only the coefficients of the components of $\mathbf{v}$ (say) need be transformed and stored, as the coefficients of the corresponding nonbasic components of $\mathbf{w}$ are immediately available. We will assume that the artificial variables are p_1 and p_2, s is the slack variable, and z the objective function. Then the initial basic feasible solution is given by

$$
\begin{aligned}
p_1 &= \quad -v_1 \qquad\qquad -2v_3 +w_1 \qquad\quad +2w_3 \\
p_2 &= \quad -v_1 -v_2 \qquad\qquad +w_1 +w_2 \\
s &= 1-v_1 -v_2 \quad -v_3 -w_1 \; -w_2 \; -w_3 \\
z &= \quad\quad v_1 +2v_2 \; +v_3 -w_1 -2w_2 \; -w_3.
\end{aligned}
$$

Then v_2 becomes basic and p_2 nonbasic, giving

$$
\begin{aligned}
p_1 &= \quad -v_1 -2v_3 +w_1 \qquad\qquad +2w_3 \\
v_2 &= \quad -v_1 \qquad\quad +w_1 +w_2 \qquad\qquad\quad -p_2 \\
s &= 1 \qquad\qquad -v_3 -2w_1 -2w_2 \; -w_3 +p_2 \\
z &= \quad -v_1 \qquad +v_3 +w_1 \qquad\qquad -w_3 -2p_2.
\end{aligned}
$$

Thus v_1 becomes basic, p_1 nonbasic, so that

$$
\begin{aligned}
v_1 &= -2v_3 +w_1 \qquad\qquad +2w_3 +p_1 \\
v_2 &= \quad 2v_3 \qquad\qquad +w_2 -2w_3 +p_1 \; -p_2 \\
s &= 1-v_3 -2w_1 -2w_2 \; -w_3 \qquad\quad +p_2 \\
z &= \quad 3v_3 \qquad\qquad\qquad -3w_3 +p_1 -2p_2.
\end{aligned}
$$

The next variable to become basic is v_3, and s must become non-basic. However, not all coefficients of v_3 in the first two equations are positive (in particular v_1 would become infeasible), so first we must interchange w_1 for v_1. This gives

$$
\begin{aligned}
w_1 &= \quad v_1 +2v_3 \qquad\qquad -2w_3 \; -p_1 \\
v_2 &= \qquad\quad 2v_3 \; +w_2 -2w_3 \; +p_1 \; -p_2 \\
s &= 1-2v_1 -5v_3 -2w_2 +3w_3 +2p_1 \; +p_2 \\
z &= \qquad\quad 3v_3 \qquad\qquad -3w_3 \; +p_1 -2p_2.
\end{aligned}
$$

Interchanging v_3 and s in the basis now gives

$$
w_1 = \frac{2}{5} - \frac{2}{5}s + \frac{1}{5}v_1 - \frac{4}{5}w_2 - \frac{4}{5}w_3 - \frac{1}{5}p_1 + \frac{2}{5}p_2
$$

$$
v_2 = \frac{2}{5} - \frac{2}{5}s - \frac{4}{5}v_1 + \frac{1}{5}w_2 - \frac{4}{5}w_3 + \frac{1}{5}p_1 - \frac{3}{5}p_2
$$

$$
v_3 = \frac{1}{5} - \frac{1}{5}s - \frac{2}{5}v_1 - \frac{2}{5}w_2 + \frac{3}{5}w_3 + \frac{2}{5}p_1 + \frac{1}{5}p_2
$$

$$
z = \frac{3}{5} - \frac{3}{5}s - \frac{6}{5}v_1 - \frac{6}{5}w_2 - \frac{6}{5}w_3 - \frac{1}{5}p_1 - \frac{7}{5}p_2.
$$

Thus the optimal solution has been obtained, with $\|\mathbf{r}\| = 3/5$ and $a_1 = 1/5$, $a_2 = 7/5$(minus the coefficients in z of p_1 and p_2 respectively).

Exercises

16. In Theorem 2.6, show that the value of j. chosen is independent of the particular vector $\boldsymbol{\mu}$ satisfying (2.11) and (2.12).

17. Use the Stiefel exchange method to solve (2.1) with

$$A = \begin{bmatrix} 1 & -1 \\ 2 & 3 \\ 3 & 1 \\ 4 & 1 \\ 5 & 0 \end{bmatrix}, \qquad b = \begin{bmatrix} 7 \\ 5 \\ 1 \\ 0 \\ 1 \end{bmatrix}$$

starting from the solution to exercise 11.

18. Prove the stated relationship between the dual basic variables in (2.13) and the vector λ of Theorem 2.1.

19. Prove that the optimal value of the slack variable required in (2.13) is zero unless $v = w = 0$ at the solution.

20. Verify the relationship between the inverse dual basis matrices referred to in this section.

21. Consider the possibility of modifying the original exchange idea so that effectively more than one index is exchanged at each step; see Hopper and Powell (1977) for a method which achieves this.

22. Consider the problem (2.1) when $m < n$ and $\| r \| = 0$ at the solution, where the particular solution with $\| a \|_\infty$ a minimum is required. Show that this can be posed and solved as a linear programming problem; see Cadzow (1973, 1974) and Ascher (1976).

2.6 Descent methods

In the previous section, it was shown how the original minimization problem defined by (2.1) could be replaced by a maximization problem, and a solution obtained to this. The direct treatment of the original problem (i.e. the primal problem in the linear programming formulation) gives rise to the class of descent methods. Although the direct solution of the primal linear programming problem as posed in the previous section is not efficient, alternative more efficient descent methods have been suggested. A recent version due to Cline (1976) which requires the matrix A to satisfy the Haar condition has been improved by Bartels, Conn, and Charalambous (1978), and we will describe the latter method.

If a does not solve (2.1), then following the proof of necessity in Theorem 2.1 there exists $c \in R^n$ such that

$$\theta_i \alpha_i^T c > 0 \qquad i \in \overline{I}(a) \tag{2.14}$$

with

$$\| r(a + \gamma c) \| < \| r(a) \|$$

for $\gamma > 0$ sufficiently small. Thus if suitable values of c and γ can be obtained, we may proceed iteratively until no vector c satisfying (2.14) exists, when the solution has been obtained. The essence of the method of Bartels *et al.* is contained in the following theorem.

Theorem 2.7 Unless **a** solves (2.1), there exists $j \in \overline{I}(\mathbf{a})$, $\delta \geqslant 0$, $\mathbf{c} \in R^n$ such that

$$\theta_i \boldsymbol{\alpha}_i^T \mathbf{c} = 1 \qquad i \in \overline{I}(\mathbf{a}) - j$$
$$\theta_j \boldsymbol{\alpha}_j^T \mathbf{c} = 1 + \delta.$$

Proof By Theorem 2.1, equation (2.14) can be satisfied if and only if the system of equations

$$\mathbf{e}^T \boldsymbol{\mu} = 1,$$

$$\sum_{i \in \overline{I}(\mathbf{a})} \mu_i \theta_i \boldsymbol{\alpha}_i = \mathbf{0} \tag{2.15}$$

$$\boldsymbol{\mu} \geqslant \mathbf{0}$$

cannot be solved. Let the system of equalities in (2.15) be written

$$G\boldsymbol{\mu} = \mathbf{e}_1 \equiv (1, 0, 0 \ldots, 0)^T, \tag{2.16}$$

where G is $(n + 1) \times k$, say (assuming that $\overline{I}(\mathbf{a}) = \{1, 2, \ldots, k\}$). Assume that the columns of G are linearly independent, and let D be a $(n + 1) \times (n + 1 - k)$ matrix whose columns are in the null space of G^T, normalized so that those columns are mutually orthonormal. Then

$$G^T D = 0 \tag{2.17}$$

$$D^T D = I. \tag{2.18}$$

Now let $\mathbf{c} \in R^n$, $\pi \in R$ be defined by

$$\begin{bmatrix} -\pi \\ \mathbf{c} \end{bmatrix} = DD^T \mathbf{e}_1. \tag{2.19}$$

If $\boldsymbol{\mu}$ solves (2.16), then

$$\begin{bmatrix} -\pi \\ \mathbf{c} \end{bmatrix} = DD^T G\boldsymbol{\mu} = \mathbf{0}.$$

Thus if $\begin{bmatrix} -\pi \\ \mathbf{c} \end{bmatrix}$ is nonzero, (2.16) cannot be solved. Further,

$$G^T \begin{bmatrix} -\pi \\ \mathbf{c} \end{bmatrix} = \mathbf{0},$$

so that

$$\theta_i \boldsymbol{\alpha}_i^T \mathbf{c} = \pi \qquad i \in \overline{I}(\mathbf{a}).$$

If $\begin{bmatrix} -\pi \\ \mathbf{c} \end{bmatrix} \neq \mathbf{0}$, π cannot be zero; further, the scaling of $\begin{bmatrix} -\pi \\ \mathbf{c} \end{bmatrix}$ can be adjusted so that $\pi = 1$. Thus the result of the theorem follows with $\delta = 0$.

Now assume that D cannot be defined, or that $\begin{bmatrix} -\pi \\ \mathbf{c} \end{bmatrix}$ defined by (2.19) is zero. Then there exists $\boldsymbol{\mu}$ satisfying (2.16). If $\boldsymbol{\mu} \geqslant \mathbf{0}$ then equation (2.14) cannot be solved and we have a solution to the problem. Otherwise determine an index j such that $\mu_j < 0$. Since the columns of G are assumed linearly independent, we can solve

$$ G^T \begin{bmatrix} -\pi \\ \mathbf{c} \end{bmatrix} = \mathbf{e}_j $$

for π and $\mathbf{c}$. Thus

$$ \theta_i \boldsymbol{\alpha}_i{}^T \mathbf{c} = \pi \qquad i \in \overline{I}(\mathbf{a}) - j $$
$$ \theta_j \boldsymbol{\alpha}_j{}^T \mathbf{c} = \pi + 1. $$

In addition

$$ \boldsymbol{\mu}^T G^T \begin{bmatrix} -\pi \\ \mathbf{c} \end{bmatrix} = \mathbf{e}_1{}^T \begin{bmatrix} -\pi \\ \mathbf{c} \end{bmatrix} = -\pi = \mu_j $$

showing that $\pi > 0$. After suitable scaling of $\mathbf{c}$ we obtain the result of the theorem, with $\delta = -1/\mu_j$. $\qquad\square$

The details contained in the proof of this theorem may be used constructively in the determination of a suitable vector $\mathbf{c}$ satisfying (2.14). If such a vector is obtained with $\delta = 0$ then

$$ r_i(\mathbf{a} + \gamma \mathbf{c}) = r_i(\mathbf{a}) - \gamma \theta_i \qquad i \in \overline{I}(\mathbf{a}) $$
$$ r_i(\mathbf{a} + \gamma \mathbf{c}) = r_i(\mathbf{a}) - \gamma \boldsymbol{\alpha}_i{}^T \mathbf{c} \qquad i \notin \overline{I}(\mathbf{a}). $$

Thus the value of $\|\mathbf{r}\|$ will be reduced to $\|\mathbf{r}(\mathbf{a})\| - \gamma$ at the point $\mathbf{a} + \gamma \mathbf{c}$, with $\overline{I}(\mathbf{a} + \gamma \mathbf{c}) \supset \overline{I}(\mathbf{a})$ for all $\gamma > 0$ sufficiently small that

$$ |r_i(\mathbf{a} + \gamma \mathbf{c})| \leqslant \|\mathbf{r}(\mathbf{a})\| - \gamma \qquad i \notin \overline{I}(\mathbf{a}). $$

We can choose the smallest positive value of γ at which

$$ |r_i(\mathbf{a} + \gamma \mathbf{c})| = \|\mathbf{r}(\mathbf{a})\| - \gamma $$

for some $i \notin \overline{I}(\mathbf{a})$ in order to define a new point $\mathbf{a} + \gamma \mathbf{c}$. If $\overline{I}(\mathbf{a})$ contains k indices, then $\overline{I}(\mathbf{a} + \gamma \mathbf{c})$ will contain at least $k + 1$ indices.

If δ in Theorem 2.7 cannot be chosen as zero, then we will have

$$ r_i(\mathbf{a} + \gamma \mathbf{c}) = r_i(\mathbf{a}) - \gamma \theta_i \qquad i \in \overline{I}(\mathbf{a}) - j, $$
$$ r_j(\mathbf{a} + \gamma \mathbf{c}) = r_j(\mathbf{a}) - \gamma \theta_j (1 + \delta), $$
$$ r_i(\mathbf{a} + \gamma \mathbf{c}) = r_i(\mathbf{a}) - \gamma \boldsymbol{\alpha}_i{}^T \mathbf{c} \qquad i \notin \overline{I}(\mathbf{a}). $$

In this case, the value of $\|\mathbf{r}\|$ will again be reduced to $\|\mathbf{r}(\mathbf{a})\| - \gamma$ at $\mathbf{a} + \gamma \mathbf{c}$ for $\gamma > 0$

sufficiently small, and attained for all $i \in \overline{I}(\mathbf{a}) - j$. The value of γ may be chosen as before. Note that $\overline{I}(\mathbf{a} + \gamma\mathbf{c})$ may have no more than k indices in this case.

Linear dependence of the columns of G corresponds to degeneracy in the linear programming formulation (see exercise 23) and may be resolved by the introduction of small perturbations in the data. For full details of the method, including the efficient and numerically stable treatment of the linear algebraic tasks involved, and some alternative ways of choosing γ, the reader is referred to the original paper (Bartels, Conn, and Charalambous, 1978). We illustrate the method as described here by a simple example.

Example $m = 3$, $n = 2$,

$$A = \begin{bmatrix} 1 & 2 \\ 1 & 1 \\ 1 & 0 \end{bmatrix}, \qquad \mathbf{b} = \begin{bmatrix} 1 \\ 2 \\ 1 \end{bmatrix}.$$

Let $\mathbf{a} = (0, 0)^T$ so that $\overline{I} = \{2\}$, $\mathbf{r} = (1, 2, 1)^T$, $\theta_2 = +1$, $\| \mathbf{r} \| = 2$.
   Then

$$G = \begin{bmatrix} 1 \\ 1 \\ 1 \end{bmatrix} \quad \text{and} \quad D = \begin{bmatrix} 1/\sqrt{2} & 1/\sqrt{6} \\ 0 & -2/\sqrt{6} \\ -1/\sqrt{2} & 1/\sqrt{6} \end{bmatrix},$$

so that

$$\begin{bmatrix} -\pi \\ \mathbf{c} \end{bmatrix} = \begin{bmatrix} \frac{2}{3} \\ -\frac{1}{3} \\ -\frac{1}{3} \end{bmatrix}.$$

Scaling so that $\pi = 1$ gives $\mathbf{c} = (\frac{1}{2}, \frac{1}{2})^T$. Now

$$|r_1(\mathbf{a} + \gamma\mathbf{c})| = |1 - \tfrac{3}{2}\gamma|,$$
$$|r_2(\mathbf{a} + \gamma\mathbf{c})| = |2 - \gamma|,$$
$$|r_3(\mathbf{a} + \gamma\mathbf{c})| = |1 - \tfrac{1}{2}\gamma|.$$

Thus we take $\gamma = 6/5$, $\mathbf{a} = (\frac{3}{5}, \frac{3}{5})^T$, $\mathbf{r} = (-4/5, 4/5, 2/5)^T$. Then $\overline{I} = \{1, 2\}$, $\theta_1 = -1$, $\theta_2 = +1$, and

$$G = \begin{bmatrix} 1 & 1 \\ -1 & 1 \\ -2 & 1 \end{bmatrix}.$$

We have

$$D = \begin{bmatrix} 1/\sqrt{14} \\ -3/\sqrt{14} \\ 2/\sqrt{14} \end{bmatrix} \quad \text{so that} \quad \begin{bmatrix} -\pi \\ \mathbf{c} \end{bmatrix} = \begin{bmatrix} 1/14 \\ -3/14 \\ 1/7 \end{bmatrix}.$$

Scaling so that $\pi = 1$ gives $\mathbf{c} = (3, -2)^T$. Now

$$|r_1(\mathbf{a} + \gamma \mathbf{c})| = |4/5 - \gamma|,$$
$$|r_2(\mathbf{a} + \gamma \mathbf{c})| = |4/5 - \gamma|,$$
$$|r_3(\mathbf{a} + \gamma \mathbf{c})| = |2/5 - 3\gamma|.$$

Thus we take $\gamma = 3/10$, $\mathbf{a} = (3/2, 0)^T$, $\mathbf{r} = (-\frac{1}{2}, \frac{1}{2}, -\frac{1}{2})^T$. Then $\overline{I} = \{1, 2, 3\}$, $\theta_1 = -1, \theta_2 = +1, \theta_3 = -1$, so that

$$G = \begin{bmatrix} 1 & 1 & 1 \\ -1 & 1 & -1 \\ -2 & 1 & 0 \end{bmatrix}.$$

The matrix D is not defined, and if we solve

$$G\mu = e_1,$$

we find that $\mu = (\frac{1}{4}, \frac{1}{2}, \frac{1}{4})$. Thus the conditions of Theorem 2.1 are satisfied, and the problem has been solved.

Exercises

23. In the proof of Theorem 2.7, it is assumed that the columns of G are linearly independent. Show that linear dependence corresponds to the failure of A to satisfy the Haar condition.

24. Prove that the method outlined above will converge in a finite number of steps to a solution of (2.1), provided that A satisfies the Haar condition.

25. Solve the problem illustrating the above method by the method of Barrodale and Phillips (1975a), and vice versa.

2.7 Some other approaches

A number of other methods for solving (2.1) have been proposed. The algorithm of Lawson (see Rice, 1969; Cline, 1972) is based on the solution of a sequence of weighted L_2 approximation problems in R^m. The method suffers from slow convergence, and in its present form does not appear to be competitive. The fact that it is relatively easy to solve problems of the type (2.1) in a smooth strictly convex L_p space is also the basis of methods which stem from an idea due to Polya in 1913. These methods make use of the following result.

Theorem 2.8 Let A have rank n, and $\mathbf{a}_p$ be the (unique) point at which $\|\mathbf{r}(\mathbf{a})\|_p$ is minimized, $1 < p < \infty$. Then if $\mathbf{a}^*$ is a limit point of $\{\mathbf{a}_p\}$ as $p \to \infty$, $\mathbf{a}^*$ solves (2.1).

Proof Let $\mathbf{r}_p = \mathbf{b} - A\mathbf{a}_p$. Then, using exercise 14 of Chapter 1, $\|\mathbf{r}_p\|_p \to \delta$, where δ is the minimum value of $\|\mathbf{r}(\mathbf{a})\|_\infty$. Let $\mathbf{a}^*$ be a limit point of $\{\mathbf{a}_p\}$ as $p \to \infty$ (which exists since the sequence is bounded), and assume that $\mathbf{a}^*$ does not solve

(2.1). Then for some $\varepsilon > 0$

$$\delta + \varepsilon \leqslant \| \mathbf{b} - A\mathbf{a}^* \|_\infty \leqslant \| \mathbf{r}_p \|_\infty + \| A(\mathbf{a}_p - \mathbf{a}^*) \|_\infty$$
$$\leqslant \| \mathbf{r}_p \|_p + \| A(\mathbf{a}_p - \mathbf{a}^*) \|_\infty .$$

We have a contradiction, and the result follows. $\qquad\square$

If (2.1) has a unique solution $\mathbf{a}^*$, then $\mathbf{a}_p \to \mathbf{a}^*$ as $p \to \infty$. Even if the solution to (2.1) is not unique, it may be shown that the sequence $\{\mathbf{a}_p\}$ is still convergent (Descloux, 1963). The limit point is shown by Descloux (1963) to give the *strict Chebyshev approximation*, introduced by Rice (1962a) as the 'best of all' best Chebyshev approximations in the following sense. If the set of solutions to (2.1) is not a singleton, then some of the components of $\mathbf{r}(\mathbf{a})$ will have degrees of freedom, as $\mathbf{a}$ ranges over all these solutions. Suppose that we require that, for these components, the maximum absolute value be minimized over all solutions. Any remaining degrees of freedom can be removed in the same way, until eventually we obtain a unique point $\mathbf{a}^*$, the strict Chebyshev solution. An exchange method for directly computing this particular solution to (2.1) is given by Duris and Temple (1973).

A recent numerical implementation of the Polya algorithm is given by Fletcher, Grant, and Hebden (1971), who use an extrapolation technique to accelerate convergence. The problem is treated in a less direct manner by Boggs (1974), who derives a differential equation describing $\mathbf{a}_p$ as a function of p. It is claimed that the method is a competitive one. Present indications are, however, that of all the methods which have currently been implemented, methods of ascent type can be generally superior.

Exercise

26. Determine the strict Chebyshev solution to the problem of exercise 3, and also to the problem defined by the first 3 components in the worked example of Section 2.4.

3

Linear Chebyshev approximation of continuous functions

3.1 Introduction

Let $f(\mathbf{x})$, $\phi_i(\mathbf{x})$, $i = 1, 2, \ldots, n$ be continuous functions of $\mathbf{x} \in X$, where X is a compact set, and let the space $C[X]$ be normed by

$$\| f(\mathbf{x}) \| = \max_{\mathbf{x} \in X} | f(\mathbf{x}) |.$$

Then we consider in this chapter the linear Chebyshev approximation problem:

$$\text{find } \mathbf{a} \in R^n \text{ to minimize } \| \mathbf{r}(\mathbf{x}, \mathbf{a}) \| \tag{3.1}$$

where

$$r(\mathbf{x}, \mathbf{a}) = f(\mathbf{x}) - \sum_{i=1}^{n} a_i \phi_i(\mathbf{x}) \qquad \mathbf{x} \in X. \tag{3.2}$$

If X is just a finite set of points, then of course this problem is identical with that treated in Chapter 2. We will assume therefore that X contains an infinite number of points, in fact that X is a *continuum* (for example the region in R^N defined by $[s_1, t_1] \times [s_2, t_2] \times \cdots \times [s_N, t_N]$), although many of the results are valid for more general regions than this. We will also assume that the functions $\{\phi_i(\mathbf{x})\}$ are linearly independent on X (for otherwise they could be replaced by a smaller number of linearly independent functions).

Even in the infinite case, the close relationship between the problems (2.1) and (3.1) is clear; if in (2.1) m is allowed to become infinite, so that we can associate a row of A (and an element of $\mathbf{b}$) with *each* point of X, then the two problems are, in a sense, equivalent. Not surprisingly, therefore, a number of concepts (and results) occurring in this chapter have precise analogues which have already appeared in

Chapter 2. For example, an important role is again played by values of $r(\mathbf{x}, \mathbf{a})$ where $\| r(\mathbf{x}, \mathbf{a}) \|$ is attained, this time defined by points $\mathbf{x} \in X$, and best approximations may again be characterized in terms of such points. This is considered formally in the next section.

3.2 Characterization of solutions

Let $\overline{E}(= \overline{E}(\mathbf{a}))$ denote the set $\mathbf{x} \in X$ for which

$$| r(\mathbf{x}, \mathbf{a}) | = \| r(\mathbf{x}, \mathbf{a}) \|.$$

The following theorem is the analogue of Theorem 2.1, and is due to Kirchberger (1903).

Theorem 3.1 A vector $\mathbf{a} \in R^n$ solves (3.1) if and only if there exists $E \subset \overline{E}$ containing $t \leqslant n + 1$ points $\mathbf{x}_1, \mathbf{x}_2, \ldots, \mathbf{x}_t$, and a nontrivial vector $\lambda \in R^t$ such that

$$\sum_{i=1}^{t} \lambda_i \phi_j(\mathbf{x}_i) = 0 \qquad j = 1, 2, \ldots, n,$$

$$\lambda_i \theta_i \geqslant 0 \qquad i = 1, 2, \ldots, t,$$

where $\theta_i = \text{sign}\,(r(\mathbf{x}_i, \mathbf{a}))$, $i = 1, 2, \ldots, t$.
(If $\| r \| = 0$ this result is trivial, so assume not.)

Proof 1 In the notation of Theorem 1.7, we have

$$V(r) = \text{conv}\{\text{sign}\,(r(\mathbf{x}, \mathbf{a}))\delta(\mathbf{x}), \mathbf{x} \in \overline{E}\}.$$

Thus $v \in V(r)$ implies that there exists $E = \{\mathbf{x}_1, \mathbf{x}_2, \ldots, \mathbf{x}_t\} \subset \overline{E}$ such that

$$v = \sum_{i=1}^{t} \mu_i \, \text{sign}\,(r(\mathbf{x}_i, \mathbf{a}))\delta(\mathbf{x}_i)$$

where $\mu_i \geqslant 0$, $i = 1, 2, \ldots, t$, $\sum_{i=1}^{t} \mu_i = 1$. The required result follows as an immediate consequence of Theorem 1.7 and Theorem 1.4.

Proof 2 Suppose that $\mathbf{a}$ does not solve (3.1) but the conditions of the theorem are satisfied. Then there exists $\mathbf{c} \in R^n$ such that

$$\| r(\mathbf{x}, \mathbf{a} + \mathbf{c}) \| < \| r(\mathbf{x}, \mathbf{a}) \|.$$

In particular

$$| r(\mathbf{x}, \mathbf{a} + \mathbf{c}) | < | r(\mathbf{x}, \mathbf{a}) | \qquad \mathbf{x} \in E,$$

or

$$\left| r(\mathbf{x}, \mathbf{a}) - \sum_{j=1}^{n} c_j \phi_j(\mathbf{x}) \right| < | r(\mathbf{x}, \mathbf{a}) | \qquad \mathbf{x} \in E,$$

so that we must have

$$\theta_i \sum_{j=1}^{n} c_j \phi_j(\mathbf{x}_i) > 0, \qquad i = 1, 2, \ldots, t.$$

Now the conditions of the theorem can be written

$$\sum_{i=1}^{t} \mu_i \theta_i \phi_j(\mathbf{x}_i) = 0 \qquad j = 1, 2, \ldots, n$$

where $\mu_i = \lambda_i \theta_i \geqslant 0, i = 1, 2, \ldots, t$. In view of the previous set of inequalities and the nontriviality of λ, this gives a contradiction.

Now let $\mathbf{a}$ be a solution to (2.1), and assume that the conditions of the theorem are not satisfied. Then

$$\mathbf{0} \notin D$$

where

$$D = \mathrm{conv}\{\mathbf{d}(\mathbf{x}), \mathbf{x} \in \overline{E}\}$$

with

$$\mathbf{d}(\mathbf{x}) = \mathrm{sign}\,(r(\mathbf{x}, \mathbf{a}))\,(\phi_1(\mathbf{x}), \phi_2(\mathbf{x}), \ldots, \phi_n(\mathbf{x}))^T.$$

Thus, by Theorem 1.5, there exists $\mathbf{c} \in R^n$ such that

$$r(\mathbf{x}, \mathbf{a}) \sum_{j=1}^{n} c_j \phi_j(\mathbf{x}) > 0 \qquad \mathbf{x} \in \overline{E}. \tag{3.3}$$

Let

$$\varepsilon = \min_{\mathbf{x} \in \overline{E}} \{ r(\mathbf{x}, \mathbf{a}) \sum_{j=1}^{n} c_j \phi_j(\mathbf{x}) \} > 0$$

and define X_1 by

$$X_1 = \{ \mathbf{x} \in X : r(\mathbf{x}, \mathbf{a}) \sum_{j=1}^{n} c_j \phi_j(\mathbf{x}) \leqslant \varepsilon/2 \}.$$

Then X_1 is a closed set containing no points of $\overline{E}$, and so $|r(\mathbf{x}, \mathbf{a})|$ achieves its maximum M on X_1, with $M < \|r(\mathbf{x}, \mathbf{a})\|$. Let $\mathbf{x} \in X_1$. Then if $\gamma > 0$,

$$\left| r(\mathbf{x}, \mathbf{a}) - \gamma \sum_{j=1}^{n} c_j \phi_j(\mathbf{x}) \right| \leqslant |r(\mathbf{x}, \mathbf{a})| + \gamma \left| \sum_{j=1}^{n} c_j \phi_j(\mathbf{x}) \right|$$

$$\leqslant M + \gamma C$$

$$< \|r(\mathbf{x}, \mathbf{a})\|$$

provided that $0 < \gamma < (\|r(\mathbf{x}, \mathbf{a})\| - M)/C = M_1$, say, where

$$C = \left\| \sum_{j=1}^{n} c_j \phi_j(\mathbf{x}) \right\|.$$

Now let $\mathbf{x} \in X - X_1$. Then

$$\left[r(\mathbf{x}, \mathbf{a}) - \gamma \sum_{j=1}^{n} c_j \phi_j(\mathbf{x})\right]^2 = [r(\mathbf{x}, \mathbf{a})]^2 - 2\gamma r(\mathbf{x}, \mathbf{a}) \sum_{j=1}^{n} c_j \phi_j(\mathbf{x})$$
$$+ \gamma^2 \left[\sum_{j=1}^{n} c_j \phi_j(\mathbf{x})\right]^2$$
$$< \|r(\mathbf{x}, \mathbf{a})\|^2 + \gamma(-\varepsilon + \gamma C^2)$$
$$< \|r(\mathbf{x}, \mathbf{a})\|^2$$

provided that $0 < \gamma < \varepsilon/C^2 = M_2$, say. Thus any choice of γ satisfying $0 < \gamma < \min(M_1, M_2)$ results in $\mathbf{a} + \gamma \mathbf{c}$ giving a better approximation than $\mathbf{a}$, a contradiction which proves the result. $\square$

Corollary 1 If $\mathbf{a}$ solves (3.1), then it also solves the L_∞ problem in R^t defined by taking $m = t$, $b_i = f(\mathbf{x}_i)$, $i = 1, 2, \ldots, t$, and A the $t \times n$ matrix with ith row $[\phi_1(\mathbf{x}_i), \ldots, \phi_n(\mathbf{x}_i)]$, $i = 1, 2, \ldots, t$, in (2.1).

Proof This is an immediate consequence of Theorem 2.1. $\square$

Corollary 2 Let $\mathbf{a}$ solve (3.1) and let E be such that $\lambda_i \neq 0, i = 1, 2, \ldots, t$. Then

$$r(\mathbf{d}, \mathbf{x}_i) = r(\mathbf{a}, \mathbf{x}_i) \qquad i = 1, 2, \ldots, t,$$

for all solutions $\mathbf{d}$ to (3.1).

Proof This is similar to that of Corollary 2 to Theorem 2.1, and is omitted. $\square$

Corollary 3 Let Y be any subset of t points from X. Then

$$\min_{\mathbf{a}} \left\{ \max_{\mathbf{x} \in Y} |r(\mathbf{x}, \mathbf{a})| \right\} \leqslant \min_{\mathbf{a}} \left\{ \max_{\mathbf{x} \in E} |r(\mathbf{x}, \mathbf{a})| \right\} = \min_{\mathbf{a}} \|r(\mathbf{x}, \mathbf{a})\|.$$

Proof This is similar to that of Theorem 2.3, and is omitted. $\square$

The last result provides the same algorithmic motivation as the corresponding result in R^m, although there is an important difference. For the problem (3.1), it does not follow that there always exists a solution with $\overline{E}$ containing $(n + 1)$ points as in the full rank case of problem (2.1).

Example $X = [0, 2]$, $f(x) = x^2$, $n = 2$, $\phi_1(x) = x$, $\phi_2(x) = e^x$.
The best approximation (correct to 4 decimal places) is

$$0.1842x + 0.4186e^x$$

with $\overline{E} = \{\xi, 2\}$ where ξ satisfies $\xi e^2 = 2e^\xi$. This is just the condition that the vector λ of Theorem 3.1 exists. The verification of these details is left as an exercise.

There is, however, an important subclass of approximation problems of the form (3.1) in which $\overline{E}$ always contains (at least) $(n + 1)$ points, and for which Theorem 3.1 may be interpreted in an extremely convenient and useful form.

Definition 3.1 The set of functions $\phi_i(\mathbf{x})$, $i = 1, 2, \ldots, n$ forms a *Chebyshev set* on X if any nontrivial linear combination has at most $(n - 1)$ zeros in X.

There is a close relationship between this condition and the Haar condition defined on a matrix: every determinant of the $n \times n$ matrix with (i, j) element $\phi_j(\mathbf{x}_i)$ for n distinct points $\mathbf{x}_i \in X$ is nonzero if and only if the functions $\{\phi_j(\mathbf{x})\}$ form a Chebyshev set on X (exercise 4). Such a set of functions is often called a *Haar system*, and the subspace of approximations a *Haar subspace*. In fact the condition is extremely restrictive, as there are no Chebyshev sets of functions of more than one variable, except for the trivial case of $n = 1$, $\phi_1(\mathbf{x}) = 1$. This was first pointed out by Mairhuber (1956).

Theorem 3.2 If $X = [a, b] \times [c, d]$, then there is no Chebyshev set of n continuous functions defined on X, for $n > 1$.

Proof Let curves α, β, and γ be as illustrated in Figure 6. For any choice of functions $\phi_1(\mathbf{x})$, $\phi_2(\mathbf{x})$, $\ldots$, $\phi_n(\mathbf{x})$ define the determinant

$$\Delta(\mathbf{x}_1, \mathbf{x}_2, \ldots, \mathbf{x}_n) = \det \begin{vmatrix} \phi_1(\mathbf{x}_1) & \cdots & \phi_n(\mathbf{x}_1) \\ \cdot & & \cdot \\ \cdot & & \cdot \\ \phi_1(\mathbf{x}_n) & \cdots & \phi_n(\mathbf{x}_n) \end{vmatrix}, \tag{3.4}$$

and choose n distinct points such that $\mathbf{x}_1$ lies on the branch α and $\mathbf{x}_2$ on γ. Now Δ is a continuous function of the points $\mathbf{x}_i$, so consider interchanging the points as follows:

(i) move $\mathbf{x}_1$ along α to $\mathbf{y}$ and then up β

(ii) move $\mathbf{x}_2$ along γ to $\mathbf{y}$ and then to where $\mathbf{x}_1$ was originally

(iii) move $\mathbf{x}_1$ down β to $\mathbf{y}$ and then along γ to where $\mathbf{x}_2$ was originally.

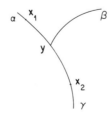

Figure 6

Clearly no two points have coincided. However, 2 rows of the determinant have been interchanged, and so its value has gone through a zero. Thus $\phi_1(\mathbf{x}), \ldots,$ $\phi_n(\mathbf{x})$ is not a Chebyshev set on X. $\qquad\square$

Despite this result, there are important examples of Chebyshev sets for the case $N = 1$, when X is the interval $[a, b]$ of the real line. The most common is the case $\phi_i(x) = x^{i-1}, i = 1, 2, \ldots, n$, where the approximation is a polynomial of degree $n - 1$; clearly no nontrivial polynomial of degree $(n - 1)$ can have more than $(n - 1)$ zeros on *any* interval. Some other examples are given in exercise 6. To motivate the particular form which Theorem 3.1 takes in the Chebyshev set case, consider the simple example of the approximation of x^2 by $a_1 + a_2 x$ in $[0, 1]$. A moment's thought and different trial placements of possible approximating lines will show that the best approximation must be as in Figure 7: it must be such that any perturbation of the line relative to the curve $y = x^2$ will result in an increase in the value of the maximum error, or in particular an increase in this value at one of the points where it is attained. Thus the line must be such that the maximum error is attained at the 3 points 0, 1, and ξ, where ξ is well-defined theoretically, but not immediately known. Notice that there are $n + 1 = 3$ points in $\overline{E}$, with the sign of the errors alternating as we move through these points from left to right: we will prove that this is a result which holds in general. We introduce the following definition, and for a proof based on Theorem 3.1 a preliminary lemma is also required.

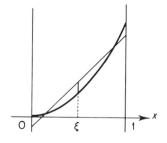

Figure 7

Definition 3.2 Let $X = [a, b]$. The function $r(x, \mathbf{a})$ is said to *alternate* s times on $[a, b]$ if $\overline{E}(\mathbf{a})$ contains $(s + 1)$ points $a \leqslant x_1 < x_2 < \ldots < x_{s+1} \leqslant b$ such that

$$r(x_i, \mathbf{a}) = -r(x_{i+1}, \mathbf{a}) \qquad i = 1, 2, \ldots, s.$$

The set $\{x_i\}$ is called an *alternating set*.

Lemma 3.1 Let $X = [a, b]$ and let the functions $\phi_i(x), i = 1, 2, \ldots, n$ form a Chebyshev set on $[a, b]$. Let $a \leqslant x_1 < x_2 < \ldots < x_{n+1} \leqslant b$ and define the $n \times n$

determinant $\Delta_i = \Delta(x_1, x_2, \ldots, x_{i-1}, x_{i+1}, \ldots, x_{n+1})$ as in equation (3.4). Then

$$\text{sign } (\Delta_i) = \text{sign } (\Delta_{i+1}) \qquad i = 1, 2, \ldots, n.$$

Proof By the Chebyshev set assumption $\Delta_i \neq 0, i = 1, 2, \ldots, n+1$. Suppose that $\Delta_j < 0 < \Delta_k$ for some j, k, with

$$\Delta_j = \Delta(y_1, y_2, \ldots, y_n),$$
$$\Delta_k = \Delta(z_1, z_2, \ldots, z_n),$$

where the order of the points in $[a, b]$ is retained. Then, for some $\gamma \in (0, 1)$,

$$\Delta(\gamma y_1 + (1-\gamma)z_1, \ldots, \gamma y_n + (1-\gamma)z_n) = 0.$$

It follows that, for some distinct l, m

$$\gamma y_l + (1-\gamma)z_l = \gamma y_m + (1-\gamma)z_m,$$

and so

$$\gamma(y_l - y_m) = (\gamma - 1)(z_l - z_m).$$

This contradicts the ordering assumption, and completes the proof. ☐

The following result was first given by Young in 1907, although the special case of polynomials was treated by Kirchberger in 1903.

Theorem 3.3 Let $X = [a, b]$ and let the functions $\phi_i(x), i = 1, 2, \ldots, n$ form a Chebyshev set on $[a, b]$. Then **a** solves (3.1) if and only if there exists an alternating set of $(n+1)$ points in $[a, b]$.

Proof 1 The Chebyshev set assumption means that $t = n+1$ in Theorem 3.1 and $\lambda_i \neq 0, i = 1, 2, \ldots, n+1$. Furthermore, λ_i is proportional to $(-1)^i \Delta_i$ (by Cramer's rule) and so, by Lemma 3.1, the components of λ alternate in sign. Thus the conditions of Theorem 3.1 and 3.2 are identical.

Proof 2 Suppose such an alternating set $\{x_i\}, i = 1, 2, \ldots, n+1$ exists, but there exists $\mathbf{c} \in R^n$ such that $\| r(x, \mathbf{c}) \| < \| r(x, \mathbf{a}) \|$. Then in particular

$$|r(x_i, \mathbf{c})| < |r(x_i, \mathbf{a})| \qquad i = 1, 2, \ldots, n+1$$

and so the difference

$$\sum_{j=1}^{n} (a_j - c_j)\phi_j(x_i)$$

alternates in sign as i goes from 1 to $n+1$. Thus there must exist n points z_1, $z_2, \ldots, z_n$ in $[a, b]$ such that

$$\sum_{j=1}^{n} (a_j - c_j)\phi_j(z_i) = 0 \qquad i, = 1, 2, \ldots, n,$$

which contradicts the Chebyshev set assumption, and shows that **a** solves (3.1).

Now let ·**a** solve (3.1) with $\| r(x, \mathbf{a}) \| = h > 0$, and assume that the largest alternating set consists of $(k + 1) < (n + 1)$ points. (If $\| r(x, \mathbf{a}) \| = 0$, the result is immediate.) Let $[a, b]$ have defined distinct subintervals $I_j, j = 1, 2, \ldots, N$ in each of which there is *one* point where h is attained (unless there is an *interval* of such points, in which case the whole interval is included), and further that

$$|r(x, \mathbf{a})| > \tfrac{1}{2}h \qquad x \in I_j.$$

Let these sub-intervals be defined from left to right (see Figure 8) and divided into subsets G_l such that each G_l contains all successive I_j in which $r(x, \mathbf{a})$ has the same sign. (For example, in the illustration, $G_1 = \{I_1\}$, $G_2 = \{I_2, I_3\}$, $G_3 = \{I_4, I_5, I_6, I_7\}$, etc.) Thus there are $(k + 1) < (n + 1)$ sets $G_l, l = 1, 2, \ldots,$ $k + 1$. Now let points $z_1, z_2, \ldots, z_k$ be defined, satisfying

$$\left. \begin{array}{ll} z_l > x & x \in G_l \\ z_l < x & x \in G_{l+1} \end{array} \right\} \quad l = 1, 2, \ldots, k,$$

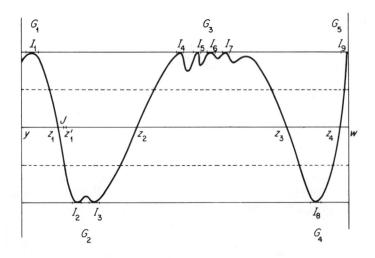

Figure 8

so that, for example, we may choose points where $r(x, \mathbf{a})$ is zero. The next part of the proof involves the construction of a function $\sum_{j=1}^{n} c_j \phi_j(x)$ which is such that it has the *same sign* as $r(x, \mathbf{a})$ in each interval $I_j, j = 1, 2, \ldots, N$. If the approximating function is a polynomial of degree $(n - 1)$, so that $\phi_j(x) = x^{j-1}$, $j = 1, 2, \ldots, n$, then we can choose (with appropriate sign)

$$\sum_{j=1}^{n} c_j \phi_j(x) = \pm \prod_{l=1}^{k} (x - z_l).$$

Otherwise such a function is constructed as follows. There exists an interval containing z_1, say J, in which no point of I_1 or I_2 lies. Let us insert points z'_1, $z'_2, \ldots, z'_s$ into J, so that $z_1 < z'_1 < \ldots < z'_s$ where $s = n - k - 1$ if $n - k - 1$ is even, $s = n - k - 2$ otherwise. Now let $\mathbf{c}$ satisfy

$$\sum_{j=1}^{n} c_j \phi_j(z_l) = 0 \qquad l = 1, 2, \ldots, k$$

$$\sum_{j=1}^{n} c_j \phi_j(z'_l) = 0 \qquad l = 1, 2, \ldots, s$$

$$\sum_{j=1}^{n} c_j \phi_j(y) = \operatorname{sign}(r(y, \mathbf{a}))$$

where y is the leftmost point of G_1, plus (if $s = n - k - 2$)

$$\sum_{j=1}^{n} c_j \phi_j(w) = \operatorname{sign}(r(w, \mathbf{a}))$$

where w is the rightmost point of G_{k+1}. There is a unique solution $\mathbf{c}$ to either of these $n \times n$ linear systems. In addition the resulting function $\sum_{j=1}^{n} c_j \phi_j(x)$ has the required form: this is immediate in the case $s = n - k - 1$, as no further zeros are possible; in the other case, an additional zero in $[y, w]$ contradicts the sign pattern.

The proof is now concluded by showing that $\mathbf{a} + \gamma \mathbf{c}$, for some $\gamma > 0$, gives a better approximation than $\mathbf{a}$. Let $K = [a, b] - \bigcup_{l=1}^{k+1} G_l$ and let $x \in K$. Then K does not contain any point where $|r(x, \mathbf{a})| = h$ and so

$$\max_{x \in K} |r(x, \mathbf{a})| = h' < h.$$

Now if $\gamma > 0$,

$$|r(x, \mathbf{a} + \gamma \mathbf{c})| = |r(x, \mathbf{a}) - \gamma \sum_{j=1}^{n} c_j \phi_j(x)|$$

$$\leqslant \|r(x, \mathbf{a})\| + \gamma \left| \sum_{j=1}^{n} c_j \phi_j(x) \right|$$

$$\leqslant h' + \gamma C$$

$$< h$$

provided that $0 < \gamma < (h - h')/C$, where $C = \left\| \sum_{j=1}^{n} c_j \phi_j(x) \right\|$. Now let $x \in \bigcup_{l=1}^{k+1} G_l$.

Then if $\gamma > 0$,

$$\left| r(x, \mathbf{a} + \gamma\mathbf{c}) \right| = \left| r(x, \mathbf{a}) - \gamma \sum_{j=1}^{n} c_j \phi_j(x) \right|$$

$$= \left| \left| r(x, \mathbf{a}) \right| - \gamma \left| \sum_{j=1}^{n} c_j \phi_j(x) \right| \right|$$

by the definition of $\mathbf{c}$. The right hand side is the modulus of a positive function provided that $0 < \gamma < h/2C$, and so

$$\left| r(x, \mathbf{a} + \gamma\mathbf{c}) \right| < h$$

provided this holds. Thus a better approximation is obtained by taking $\mathbf{a} + \gamma\mathbf{c}$ where $0 < C\gamma < \min\{h - h', h/2\}$, giving a contradiction, and concluding the proof. $\qquad\square$

If $\phi_1(x), \ldots, \phi_n(x)$ is *not* a Chebyshev set on X, there may or may not be a solution $\mathbf{a}$ to (3.1) for which an alternating set of $(n + 1)$ points exists. Jones and Karlowitz (1970) show that there exists a solution with such an alternating set if and only if $\phi_1(x), \ldots, \phi_n(x)$ is a *weak Chebyshev set* on X, i.e. any linear combination has at most $(n - 1)$ changes of sign.

Exercises

1. If $\mathbf{a}$ solves (3.1), prove that $\overline{E}(\mathbf{a})$ contains at least 2 points.

2. Prove that $\mathbf{a}$ solves (3.1) if and only if

$$\min_{\mathbf{x} \in \overline{E}(\mathbf{a})} r(\mathbf{x}, \mathbf{a}) \sum_{i=1}^{n} c_i \phi_i(\mathbf{x}) \leqslant 0$$

for all $\mathbf{c} \in R^n$. (This is the Kolmogoroff criterion.)

3. Let $f(x) \in C[a, b]$ be nonzero, and let $g(x) \in C[a, b]$ be such that

$$f(x)g(x) < 0$$

for all points x where $|f(x)| = \|f\|$ in $[a, b]$. Prove that there exists $\gamma > 0$ such that

$$\|f - \gamma g\| < \|f\|.$$

4. Prove that a necessary and sufficient condition for any nontrivial linear combination of the n functions $\phi_1(x), \ldots, \phi_n(x)$ to have at most $(n - 1)$ zeros in X is that the determinant of the $n \times n$ matrix with (i, j) element $\phi_j(\mathbf{x}_i)$ be nonzero for any n distinct points $\mathbf{x}_1, \mathbf{x}_2, \ldots, \mathbf{x}_n$ in X.

5. Let $\phi_1(x), \ldots, \phi_n(x)$ be a Chebyshev set on $[a, b]$. Given points $x_1 < x_2 < \ldots < x_{n-1}$ in $[a, b]$, prove that there exists $\mathbf{c} \in R^n$ such that $\sum_{i=1}^{n} c_i \phi_i(x)$ changes sign at these points and has no other zeros in $[a, b]$.

6. Prove that the following form Chebyshev sets
   (i) $\{1, x^2, x^4\}$ on $[0, 1]$
   (ii) $\{1, e^x, e^{2x}\}$ on $[0, 1]$
   (iii) $\{1, e^x, e^{2x}, \ldots, e^{(n-1)x}\}$ on $[0, 1]$
   (iv) $\{\sin x, \sin 2x, \ldots, \sin nx\}$ on $(0, \pi)$
   (v) $\{1, \sin x, \ldots, \sin nx, \cos x, \ldots, \cos nx\}$ on $[0, 2\pi)$.

7. Show that the following do not form Chebyshev sets
   (i) $\{1, x^2, x^4\}$ on $[-1, 1]$
   (ii) $\{1, x^2, x^3\}$ on $[-1, 1]$
   (iii) $\{x, e^x\}$ on $[0, 2]$
   (iv) $\{1 - 8x, 2x + x^2\}$ on $[0, 1]$.

8. Let $\phi_i(x) = x^{i-1}$, $i = 1, 2, \ldots, n-1$, $\phi_n(x) = f(x) \in C^{n-1}[a, b]$. If $D^{n-1}f(x) > 0$ in $[a, b]$, show that $\phi_i(x)$, $i = 1, 2, \ldots, n$ is a Chebyshev set on $[a, b]$. (Hint: Use Rolle's theorem.)

9. Let $f(x) \in C^2[a, b]$ and let $D^2 f(x) > 0$ in $[a, b]$. Find the best Chebyshev approximation to $f(x)$ on $[a, b]$ by a straight line. Hence find the best Chebyshev approximation to x^m on $[0, 1]$ by a straight line. Describe what happens as $m \to 0$ and as $m \to \infty$.

10. Find the best Chebyshev approximation to x^2 on $[0, 1]$ by $a_1 x$.

11. Consider the problem of determining the best Chebyshev approximation to x^m where m is an integer $\geqslant 3$ on $[0, 1]$ by a polynomial of degree 2. Prove that the best approximation alternates exactly 4 times at the points $\{0, x_1, x_2, 1\}$ with $x_1 < x_2$ and establish the following properties
   (i) $x_1 < \hat{x} < x_2$ where $\hat{x} = (2/m)^{1/(m-2)} x_2$ is a root of $D^2 r(x, \mathbf{a}) = 0$, where $\mathbf{a}$ solves the problem.
   (ii) $(m-2)x_2 < 2(m-1)x_1 < mx_2$
   (iii) $x_2 \to 1$ as $m \to \infty$, and, for large m,
   $$x_1 = \frac{1}{2} \frac{m-2}{m-1} x_2 + O(2^{-m})$$
   $$a = (1 - x_1)^{-2} + O(2^{-m})$$
   $$b = -2x_1(1 - x_1)^{-2} + O(2^{-m})$$
   $$\|r\| = \tfrac{1}{2}x_1^2 (1 - x_1)^{-2} + O(2^{-m-1}).$$
Describe the behaviour of the approximation as m becomes large.

12. Let X be the disc in R^2 defined by $x^2 + y^2 \leqslant 1$. Show that the best approximation to $f(x) = x^2 y$ by an expression of the form

$$a_1 + a_2 x + a_3 x^2 + a_4 y + a_5 y^2 + a_6 xy$$

is given by the continuum

$$0.25y + t(1 - x^2 - y^2) \qquad -0.125 \leqslant t \leqslant 0.125.$$

(Gearhart, 1973).

13.   If X contains a curve homeomorphic to a circle (and another point), show that there is no Chebyshev set $\phi_i(x)$, $i = 1, 2, \ldots, n$ of continuous functions defined on X for n odd, $n > 1$ (n even).

14.   Let $\phi_1(x), \ldots, \phi_n(x)$ form a Chebyshev set on $[a', b']$ where $a' < a < b < b'$. Given $s \leqslant n - 1$ distinct points $x_1, x_2, \ldots, x_s$ in $[a, b]$, prove that there exists $\mathbf{c} \in R^n$ such that $\sum_{i=1}^{n} c_i \phi_i(x)$ changes sign at these points and has no other zeros in $[a, b]$. (Hint: use induction.)

15.   Let $\phi_i(x), i = 1, 2, \ldots, n$ form a Chebyshev set on $[a, b]$ and for some $\mathbf{d} \in R^n$ let there exist $(n + 1)$ points $x_1, x_2, \ldots, x_{n+1}$ in $[a, b]$ such that
  (i) $r(x_i, \mathbf{d}) \neq 0$ $i = 1, 2, \ldots, n + 1$,
  (ii) sign $r(x_{i+1}, \mathbf{d}) = -\text{sign } r(x_i, \mathbf{d})$ $i = 1, 2, \ldots, n$.
Prove that

$$\min_{1 \leqslant i \leqslant n+1} |r(x_i, \mathbf{d})| \leqslant \min_{\mathbf{a}} \|r(x, \mathbf{a})\|$$

(de la Vallée Poussin, 1919).

16.   Let $\phi_i(x) = x^{i-1}, i = 1, 2, \ldots, n$. Show that the conditions (i) and (ii) of the previous exercise are sufficient for

$$\min_{1 \leqslant i \leqslant n} \tfrac{1}{2}(|r(x_i, \mathbf{d})| + |r(x_{i+1}, \mathbf{d})|) \leqslant \min_{\mathbf{a}} \|r(x, \mathbf{a})\|$$

(Remes, 1934a; see, for example Meinardus, 1967. For analogous lower estimates based on k points, $1 \leqslant k \leqslant n$, see Meinardus and Taylor, 1976).

17.   Let $\phi_1(x), \ldots, \phi_n(x)$ form a weak Chebyshev set on $[a, b]$. Given points $x_1 < x_2 < \ldots < x_{n-1}$ in $[a, b]$, prove that there exists a nontrivial vector $\mathbf{c} \in R^n$ such that

$$(-1)^{i+1} \sum_{j=1}^{n} c_j \phi_j(x) \geqslant 0 \qquad x \in (x_{i-1}, x_i), i = 1, 2, \ldots, n,$$

where $x_0 \equiv a$, $x_n \equiv b$.

18.   Let x_+^n be defined by

$$x_+^n = \begin{cases} x^n & \text{if } x \geqslant 0 \\ 0 & \text{if } x < 0 \end{cases}$$

for $x \in [a, b]$. Prove that the following is a weak Chebyshev set, where $t_1, t_2, \ldots, t_k$ are distinct points in $[a, b]$:

$$\{1, x, (x - t_1)_+, (x - t_2)_+, \ldots, (x - t_k)_+\} \text{ on } [a, b].$$

  The following are (more general) weak Chebyshev sets (see Karlin and Studden, 1966, for these and many other properties of Chebyshev and weak Chebyshev sets).

(i) $\{(x-t_1)_+{}^n, (x-t_2)_+{}^n, \ldots, (x-t_k)_+{}^n\}$ on $[-1, 1]$

(ii) $\{1, x, x^2, \ldots, x^n, (x-t_1)_+{}^n, \ldots, (x-t^k)_+{}^n\}$ on $[-1, 1]$.

19. The functions $\phi_1(x), \ldots, \phi_n(x)$ form a *Markov set* if $\phi_1(x), \ldots, \phi_j(x)$ is a Chebyshev set for $j = 1, 2, \ldots, n$. Show that $\{e^{\alpha_1 x}, e^{\alpha_2 x}, \ldots, e^{\alpha_n x}\}$ forms a Markov set on any interval. For given $X = [a, b]$, and given n, show that there exists a Chebyshev set which is not (and is not a rearrangement of) a Markov set.

20. Let $\phi_1(x), \phi_2(x), \ldots$ be an *infinite* Markov set on $[a, b]$. Prove that there does not exist a best approximation of the form $\Sigma a_i \phi_i(x)$ to any element of $C[a, b]$ not already expressible in this form. (Cheney, 1966).

21. Can you see any way of providing alternation type concepts (and thus results) for multivariate functions, or for univariate functions when X is not an interval? Look in Buck (1968), Weinstein (1968).

22. Consider the problem (3.1) with $X = [a, b]$ subject to constraints

$$Ca = d$$

where C is an $m \times n$ matrix of rank m which (without loss of generality) may be written as $[G\,H]$ where G is $m \times m$ nonsingular. If the constraints are used to eliminate $a_1, a_2, \ldots, a_m$, we obtain a corresponding unconstrained problem with $(n-m)$ approximating functions $\psi_1(x), \ldots, \psi_{n-m}(x)$. Let T be a $t \times (n-m)$ matrix with (j, i) component $\psi_i(x_j)$. If $S = [U\,V]$ is a $t \times n$ matrix with (j, i) component $\phi_i(x_j)$ where U is $t \times m$, show that we can write

$$T = V - UG^{-1}H.$$

Hence prove that necessary and sufficient conditions for **a** to solve the constrained problem are

(i) $Ca = d$

(ii) there exists $t \leqslant n + 1 - m$ points x_i, $i = 1, 2, \ldots, t$ in $\overline{E}(\mathbf{a})$, a corresponding matrix S as defined above, and a nontrivial vector $\lambda \in R^{t+m}$ such that

$$[S^T\,C^T]\lambda = 0$$

$$\lambda_1 \theta_i \geqslant 0 \qquad i = 1, 2, \ldots, t,$$

where $\theta_i = \text{sign}(r(x_i, \mathbf{a}))$, $i = 1, 2, \ldots, t$.

If the functions $\{\phi_i(x)\}$ form a Chebyshev set on $[a, b]$, and the constraints are interpolatory, so that they become

$$r(y_i, \mathbf{a}) = 0 \qquad i = 1, 2, \ldots, m$$

for points $y_1, y_2, \ldots, y_m$ in $[a, b]$, deduce that $t = n + 1 - m$ and

$$\theta_i = (-1)^{d_i+1}\theta_{i+1} \qquad i = 1, 2, \ldots, t-1,$$

where there are d_i interpolation points between x_i and x_{i+1}.

23. Consider the problem (3.1) with $X = [a, b]$ subject to the constraint

$$p(x, \mathbf{a}) = \sum_{i=1}^{n} a_i \psi_i(x) + \psi_o(x) \geq 0 \qquad x \in X,$$

where $\psi_0(x), \ldots, \psi_n(x) \in C[X]$. Let $\overline{Z}(\mathbf{a})$ be defined by

$$\overline{Z}(\mathbf{a}) = \{x \in X : p(x, \mathbf{a}) = 0\}.$$

Prove that $\mathbf{a}$ solves this constrained problem if
(i) $p(x, \mathbf{a}) \geq 0, \ x \in X,$
(ii) there exist $t \leq n + 1$ points $x_1, x_2, \ldots, x_s \in \overline{E}(\mathbf{a})$, and $x_{s+1}, \ldots, x_t \in \overline{Z}(\mathbf{a})$, and a vector $\lambda \in R^t$ such that

$$\sum_{i=1}^{s} \lambda_i \phi_j(x_i) + \sum_{i=s+1}^{t} \lambda_i \psi_j(x_i) = 0 \quad j = 1, 2, \ldots, n,$$

$$\lambda_i \theta_i \geq 0 \qquad i = 1, 2, \ldots, t,$$

$$\sum_{i=1}^{s} |\lambda_i| > 0,$$

where

$$\theta_i = \text{sign}(r(x_i, \mathbf{a})) \qquad i = 1, 2, \ldots, s,$$
$$\theta_i = 1 \qquad\qquad\quad i = s + 1, \ldots, t.$$

Prove that these conditions are necessary for $\mathbf{a}$ to solve the constrained problem, provided that $\sum_{i=1}^{s} |\lambda_i| > 0$ is replaced by $\sum_{i=1}^{t} |\lambda_i| > 0$.

24. The results of exercise 23 generalize in an obvious way to the case of constraints

$$p_j(x, \mathbf{a}) \geq 0 \qquad j = 1, 2, \ldots, r, \qquad x \in X.$$

An important special case is when $r = 2$ and

$$p_1(x, \mathbf{a}) = \sum_{i=1}^{n} a_i \phi_i(x) - l(x)$$

$$p_2(x, \mathbf{a}) = -\sum_{i=1}^{n} a_i \phi_i(x) + u(x),$$

when we have *restricted range approximation*. Show that in this case, the necessary and sufficient conditions are identical. That this is not true in general, however, may be shown for the problem on $[0, 1]$:

$$\text{find } \mathbf{a} \in R^2 \text{ to minimize } \| x^2 - a_1 - a_2 x \|$$
$$\text{subject to} \quad a_1 + a_2 x \quad -x \geq 0$$
$$\qquad\qquad -a_1 + a_2(x - 1) + (1 + x) \geq 0$$

by considering the point $\mathbf{a} = (1, 0)^T$. (There is a large body of literature on

Chebyshev approximation subject to constraints; the interested reader should refer in the first instance to the review papers by Taylor, 1973, and Chalmers and Taylor, 1978: the latter contains 168 references).

3.3 Uniqueness of the solution

The close relationship between Chebyshev sets and the Haar condition suggests that uniqueness in the present case is a likely consequence of approximations being required from a Haar subspace. This is the substance of the following theorem, due to Young (1907).

Theorem 3.4 Let $X = [a, b]$ and let the functions $\phi_i(x), i = 1, 2, \ldots, n$ form a Chebyshev set on $[a, b]$. Then the solution to (3.1) is unique.

Proof Let $\mathbf{a}$ and $\mathbf{c}$ be solutions to (3.1). Then, using convexity, $(\mathbf{a} + \mathbf{c})/2$ is also a solution. Thus by Theorem 3.3, there exist points $x_1 < x_2 < \ldots < x_{n+1}$ in $[a, b]$ such that

$$f(x_i) - \sum_{j=1}^{n} \frac{a_j + c_j}{2} \phi_j(x_i) = (-1)^i h \qquad i = 1, 2, \ldots, n+1$$

where $h = \pm \| r(x, \mathbf{a}) \|$. Thus

$$\tfrac{1}{2}[f(x_i) - \sum_{j=1}^{n} a_j \phi_j(x_i)] + \tfrac{1}{2}[f(x_i) - \sum_{j=1}^{n} c_j \phi_j(x_i)] = (-1)^i h$$
$$i = 1, 2, \ldots, n+1,$$

showing that

$$f(x_i) - \sum_{j=1}^{n} a_j \phi_j(x_i) = f(x_i) - \sum_{j=1}^{n} c_j \phi_j(x_i) = (-1)^i h \qquad i = 1, 2, \ldots, n+1.$$

Thus

$$\sum_{j=1}^{n} (a_j - c_j) \phi_j(x_i) = 0 \qquad i = 1, 2, \ldots, n+1,$$

showing that $\mathbf{a} = \mathbf{c}$ by the Chebyshev set assumption. $\qquad\square$

It is easy to construct examples of non-uniqueness in the absence of Chebyshev sets.

Example $X = [0, 1], f(x) = 1, n = 1, \phi_1(x) = x$.
The best approximation is given by $a_1 x$, for any a_1 such that $0 \leqslant a_1 \leqslant 2$.

It is also easy to construct examples of problems without Chebyshev sets which *have* unique solutions (see exercise 10). However, the following result of Haar (1918) shows that the Chebyshev set assumption is in a sense a minimal one.

Theorem 3.5 Let $\phi_1(x), \ldots, \phi_n(x)$ be given, and let the solution to (3.1) be unique for *all possible* $f(x) \in C[X]$. Then $\{\phi_i(x)\}, i = 1, 2, \ldots, n$ is a Chebyshev set on X.

Proof Suppose that the uniqueness assumption holds, but the set of functions does not form a Chebyshev set. Then by exercise 4, there exist points x_1, $x_2, \ldots, x_n$ in X such that

$$\Delta(x_1, x_2, \ldots, x_n) = 0$$

(see equation (3.4)) and so the $n \times n$ matrix M whose determinant this is must be singular. Thus there exist nontrivial vectors c and d in R^n such that

$$M^T c = M d = 0,$$

so that

$$\sum_{i=1}^{n} c_i \phi_j(x_i) = 0 \qquad j = 1, 2, \ldots, n, \tag{3.5}$$

$$g(x_i) \equiv \sum_{j=1}^{n} d_j \phi_j(x_i) = 0 \qquad i = 1, 2, \ldots, n.$$

Further, we may take $\| g(x) \| < 1$. Now let $f_1(x) \in C[X]$ be such that $\| f_1(x) \| = 1$ and

$$f_1(x_i) = \text{sign}(c_i) \qquad i = 1, 2, \ldots, n,$$

and define $f(x)$ by

$$f(x) = f_1(x)(1 - |g(x)|).$$

Then

$$f(x_i) = f_1(x_i) = \text{sign}(c_i) \qquad i = 1, 2, \ldots, n.$$

Let $a \in R^n$ be arbitrary. Then

$$\| r(x, a) \| \geqslant 1$$

for if not

$$\text{sign}\left(\sum_{j=1}^{n} a_j \phi_j(x_i) \right) = \text{sign}(f(x_i)) \qquad i = 1, 2, \ldots, n,$$

$$= \text{sign}(c_i) \qquad i = 1, 2, \ldots, n,$$

contradicting (3.5). It remains to show that for γ, $0 \leqslant \gamma \leqslant 1$, $\gamma g(x)$ is a best approximation to $f(x)$. For if $0 \leqslant \gamma \leqslant 1$, $x \in X$ is arbitrary,

$$|r(x, \gamma d)| \leqslant |f(x)| + \gamma |g(x)|$$

$$= |f_1(x)|(1 - |g(x)|) + \gamma |g(x)|$$

$$\leqslant 1 - |g(x)| + \gamma |g(x)|$$

$$\leqslant 1.$$

Thus equality must hold throughout, and the result is proved. □

As in the case of (2.1), it is possible to give some information on how fast $\|r(\mathbf{x}, \mathbf{a})\|$ increases as $\mathbf{a}$ moves away from the (unique) solution to (3.1). The following definition is just the function space analogue of Definition 2.2, and the theorem (due to Newman and Shapiro, 1962) may be proved in a similar manner to Theorem 2.4.

Definition 3.3 The solution $\mathbf{a}$ to (3.1) is *strongly unique* if there exists $\gamma > 0$ such that

$$\|r(\mathbf{x}, \mathbf{c})\| \geq \|r(\mathbf{x}, \mathbf{a})\| + \gamma \|\mathbf{a} - \mathbf{c}\|_2 \tag{3.6}$$

for all $\mathbf{c} \in R^n$ where $\|.\|_2$ is the L_2 norm on R^n.

Theorem 3.6 Let $\phi_i(x)$, $i = 1, 2, \ldots, n$ form a Chebyshev set on $X = [a, b]$. Then the solution to (3.1) is strongly unique.

Unlike the situation in R^m, uniqueness of the solution to (3.1) is not sufficient to guarantee strong uniqueness. In fact McLaughlin and Somers (1975) show that every element of $C[a, b]$ which possesses a unique best Chebyshev approximation by a linear combination of the functions $\phi_1(x), \ldots, \phi_n(x)$ also possesses a strongly unique best approximation if and only if the functions form a Chebyshev set on $[a, b]$.

Example $X = [0, 2]$, $n = 1$, $\phi_1(x) = x$

$$f(x) = \begin{cases} 1 - x^2 & 0 \leq x \leq 1 \\ 1 - x & 1 \leq x \leq 2. \end{cases}$$

The unique solution is given by $a_1 = 0$ with $\|r\| = 1$. Now

$$\|r(x, a)\| = \max \left\{ \max_{0 \leq x \leq 1} |ax - 1 + x^2|, \ \max_{1 \leq x \leq 2} |ax - 1 + x| \right\}$$

$$= \begin{cases} 1 + \dfrac{a^2}{4} & a < 0 \\ 1 + 2a & a > 0 \end{cases}$$

provided that $|a|$ remains sufficiently small. Thus for $a < 0$ small enough

$$\|r(x, a)\| - 1 = \frac{a^2}{4} \geq \gamma |a| \text{ only if } \gamma = 0.$$

Thus the solution is not strongly unique.

Exercises

25. Prove Theorem 3.6.

26. Let $\phi_i(x)$, $i = 1, 2, \ldots, n$ form a Chebyshev set on $[a, b]$, and let $\mathbf{a} \in R^n$ and $\mathbf{b} \in R^n$ solve (3.1) with $f(x) = f_1(x)$ and $f(x) = f_2(x)$ respectively. Prove that there exists a constant K depending on $f_1(x)$ such that, for all $f_2(x) \in C[a, b]$,

$$\| \mathbf{a} - \mathbf{b} \|_2 \leqslant K \| f_1(x) - f_2(x) \|.$$

(Freud, 1958.)

27. Let

$$\delta = \min_{\| \mathbf{d} \|_2 = 1} \left\| \sum_{i=1}^{n} d_i \phi_i(x) \right\|.$$

Show that γ defined by (3.6) satisfies $\gamma \leqslant \delta$.

28. Let $X = [0, 1] \times [0, 1]$, let $n \leqslant 3$, and let $f(\mathbf{x})$, $\phi_i(\mathbf{x})$, $i = 1, 2, \ldots, n$ have continuous first partial derivatives of all variables in X. Then if

(i) the $n \times 3$ matrix

$$\begin{bmatrix} \phi_1 & \dfrac{\partial \phi_1}{\partial x_1} & \dfrac{\partial \phi_1}{\partial x_2} \\ \vdots & \vdots & \vdots \\ \phi_n & \dfrac{\partial \phi_n}{\partial x_1} & \dfrac{\partial \phi_n}{\partial x_2} \end{bmatrix}$$

has rank n for any $\mathbf{x} \in (0, 1) \times (0, 1)$,

(ii) for fixed x_1 or x_2 $\phi_i(\mathbf{x})$, $i = 1, 2, \ldots, n$ forms a Chebyshev set on $[0, 1]$, prove that the solution to (3.1) is unique (see Rivlin and Shapiro, 1960, for a more general result; necessary and sufficient conditions for the uniqueness of best Chebyshev approximations for certain non-Haar subspaces are given by Weinstein, 1968).

29. Let $X = [a, b] \times [c, d]$, and let $f(\mathbf{x}) \in C^1[X]$. Prove that the best Chebyshev approximation to $f(\mathbf{x})$ by functions of the form $a_1 + a_2 x_1 + a_3 x_2$ is unique. (Collatz, 1956.)

30. Theorems 3.2 and 3.3 do not hold for approximation on a *noncompact* interval. Show this by considering the example

$$X = [0, \infty), \quad n = 2, \quad \phi_1(x) = 1, \quad \phi_2(x) = e^{-x}$$
$$f(x) = (1 - e^{-x}) \sin x.$$

(However, the alternation concept can be extended by the introduction of 'essential' alternation; see Kammler, 1976.)

31. Let $X = [a, b]$ and let $\mathbf{a} \in R^n$. Then $\gamma \geqslant 0$ satisfies (3.6) for all $\mathbf{c} \in R^n$ if and only if

$$\max_{x \in \overline{E}(\mathbf{a})} r(x, \mathbf{a}) \sum_{j=1}^{n} c_j \phi_j(x) \geqslant \gamma \| r(x, \mathbf{a}) \| \left\| \sum_{j=1}^{n} c_j \phi_j(x) \right\|$$

for all $c \in R^n$ (Bartelt and McLaughlin, 1973).

32. Let a solve either of the equality or inequality constrained problems introduced in exercises 22 and 23. If $x \in \bar{E}(a)$ occurs in the appropriate characterization result, and the corresponding component of λ is nonzero, prove that all solutions take the same value at x.

If $f(x) \in C^1(a, b)$, $\phi_i(x) = x^{i-1}$, $i = 1, 2, \ldots, n$, deduce that if $k = 0$, 1, or 2 according to the number of points $\{a, b\}$ represented in the characterization result with a corresponding nonzero component of λ, and l is the number of internal points of (a, b) so represented, then a is unique if $2l + k \geqslant n$.

33. Show that the problem in $[0, 1]$:

$$\text{find } a \in R^3 \text{ to minimize } \| (1 - 2x)(1 - x + x^2) - a_1 - a_2 x - a_3 x^2 \|$$
$$\text{subject to } a_2 + a_3 = 0$$

is solved by $a = (0, t, -t)^T$, $0 \leqslant t \leqslant 1$.

3.4 Discretization

In contrast to the methods of the previous chapter, we can not generally expect to be able to obtain a solution to (3.1) in a finite number of steps. If, however, the region X is *replaced* by a discrete set of m points in X, then the continuous problem (3.1) is replaced by a *corresponding discrete problem*

$$\text{find } a \in R^n \text{ to minimize } \max_{1 \leqslant i \leqslant m} |r(x_i, a)|. \tag{3.7}$$

This is clearly a particular example of the class of problem treated in Chapter 2, and is therefore subject to the analysis given there, and may be solved by any of the appropriate algorithms. Indeed, problems of this type play a fundamental role in algorithms for solving (3.1), as most popular methods involve the solution of a *sequence* of such problems, and differ in the way in which the discrete point sets are defined at each iteration. If the discrete set 'fills out' the region X, it may be expected that a solution to (3.7) will give a good approximation to a solution of (3.1). We will make this more precise.

Definition 3.4 Let $Y = \{x_j\} \in X$. Then the density of Y in X is defined by

$$|Y| = \max_{x \in X} \inf_{y \in Y} \| y - x \|_2$$

where the norm on R^N is the L_2 norm.

Now let $M = \min_a \| r(x, a) \|$ and let

$$M(Y) = \min_a \max_{y \in Y} |r(y, a)| = \max_{y \in Y} |r(y, a(Y))|.$$

Theorem 3.7 Let $|Y| \to 0$. Then
  (i) $M(Y) \to M$
  (ii) $\| r(\mathbf{x}, \mathbf{a}(Y)) \| \to M$
  (iii) If $\mathbf{a}^*$ is the unique solution to (3.1), $\mathbf{a}(Y) \to \mathbf{a}^*$.

Proof For given $\varepsilon > 0$, let us define

$$\omega(\varepsilon) = \sup_{\| \mathbf{x} - \mathbf{y} \|_2 \leqslant \varepsilon} |f(\mathbf{x}) - f(\mathbf{y})|$$

$$\Omega(\varepsilon) = \max_{1 \leqslant i \leqslant n} \sup_{\| \mathbf{x} - \mathbf{y} \|_2 \leqslant \varepsilon} |\phi_i(\mathbf{x}) - \phi_i(\mathbf{y})|.$$

Then, by continuity, both $\omega(\varepsilon)$ and $\Omega(\varepsilon)$ tend to zero with ε. Let ε be such that $\Omega(\varepsilon) < \delta$, where $\delta > 0$ satisfies

$$\left\| \sum_{j=1}^{n} c_j \phi_j(\mathbf{x}) \right\| \geqslant \delta \sum_{j=1}^{n} |c_j|$$

for all $\mathbf{c} \in R^n$ (using the linear independence assumption). For given $|Y| < \varepsilon$, and a corresponding $\mathbf{a}(Y)$, let $\boldsymbol{\xi} \in X$ be such that $\| r(\mathbf{x}, \mathbf{a}(Y)) \| = |r(\boldsymbol{\xi}, \mathbf{a}(Y))|$ and let $\mathbf{y} \in Y$ satisfy $\| \boldsymbol{\xi} - \mathbf{y} \|_2 \leqslant \varepsilon$.

Then

$$\begin{aligned}
M &\leqslant \| r(\mathbf{x}, \mathbf{a}(Y)) \| = |r(\boldsymbol{\xi}, \mathbf{a}(Y))| \\
&\leqslant |f(\boldsymbol{\xi}) - f(\mathbf{y})| + |r(\mathbf{y}, \mathbf{a}(Y))| \\
&\quad + \left| \sum_{j=1}^{n} a_j(Y)(\phi_j(\mathbf{y}) - \phi_j(\boldsymbol{\xi})) \right| \\
&\leqslant \omega(\varepsilon) + M(Y) + \Omega(\varepsilon) \sum_{j=1}^{n} |a_j(Y)|. \tag{3.8}
\end{aligned}$$

Now

$$\begin{aligned}
\sum_{j=1}^{n} |a_j(Y)| &\leqslant \frac{1}{\delta} \left\| \sum_{j=1}^{n} a_j(Y) \phi_j(\mathbf{x}) \right\| \\
&\leqslant \frac{2}{\delta} \max_{\mathbf{y} \in Y} \left| \sum_{j=1}^{n} a_j(Y) \phi_j(\mathbf{y}) \right| \qquad \text{using exercise 36} \\
&\leqslant \frac{2}{\delta} \left(M(Y) + \max_{\mathbf{y} \in Y} |f(\mathbf{y})| \right) \\
&\leqslant \frac{4}{\delta} \| f \|. \tag{3.9}
\end{aligned}$$

Thus

$$0 \leqslant M - M(Y) \leqslant \omega(\varepsilon) + \frac{4}{\delta} \| f \| \Omega(\varepsilon)$$

and we have (i). The inequalities to (3.8) then give (ii), as the inequalities to (3.9) show that $\{\mathbf{a}(Y)\}$ is bounded. The limit points, by (ii), must solve (3.1), and the result (iii) follows. $\qquad\square$

Theorem 3.7 shows that one way of solving (3.1) is to compute a sequence of approximations on discrete subsets of X, these subsets being such that X is more and more 'filled out'. A process such as this would, however, be extremely inefficient, and we turn now to methods for solving (3.1) which involve the solution of a sequence of problems (3.7) constructed in a much more systematic fashion.

Exercises

34. Let $X = X_1 \times X_2 \times \ldots \times X_N$, where $X_i = [0, 1]$, $i = 1, 2, \ldots, N$. If $Y \subset X$ consists of $(m+1)^N$ equispaced points, including boundary points, determine $|Y|$.

35. Let $f(\mathbf{x}) \in C[X]$ not be identically zero, and let $\varepsilon > 0$ be such that $\omega(\varepsilon) < \|f(\mathbf{x})\|$. Prove that if $|Y| < \varepsilon$,

$$\|f\| \leqslant \left[1 + \frac{\omega(\varepsilon)}{\|f\| - \omega(\varepsilon)}\right] \max_{\mathbf{y} \in Y} |f(\mathbf{y})|.$$

36. Let $\Omega(\varepsilon) < \delta, |Y| < \varepsilon$. Modify the method used in exercise 35 to prove that

$$\left\|\sum_{j=1}^{n} c_j \phi_j(\mathbf{x})\right\| \leqslant \left[1 + \frac{\Omega(\varepsilon)}{\delta - \Omega(\varepsilon)}\right] \max_{\mathbf{y} \in Y} \left|\sum_{j=1}^{n} c_j \phi_j(\mathbf{y})\right|,$$

for any $\mathbf{c} \in R^n$.

37. If $f(x)$ has a unique best approximation, there may exist finite sets $Y \subset X$ with $|Y|$ arbitrarily small such that the discrete solution on Y is non-unique. Consider the example on $X = [-1, 1]$ with $n = 1$ and

$$f(x) = \begin{cases} 1 + x & -1 \leqslant x \leqslant 0 \\ \cos(2\pi x) & 0 \leqslant x \leqslant 1 \end{cases}$$

$$\phi_1(x) = \begin{cases} 0 & -1 \leqslant x \leqslant 0 \\ x & 0 \leqslant x \leqslant 1 \end{cases}$$

for which the unique solution is $a_1 = 0$ (Dunham, 1975).

3.5 The algorithms of Remes

Two algorithms for solving (3.1) were given by Remes in the 1930s. We begin by describing that traditionally known as the 'First Algorithm', which applies to general problems of the class. At the kth step is defined a discrete subset X^k of X. A vector $\mathbf{a}^k \in R^n$ is obtained minimising $\max_{\mathbf{x} \in X^k} |r(\mathbf{x}, \mathbf{a})|$ and a point $\mathbf{x}^k \in X$ at which

$|r(\mathbf{x}, \mathbf{a}^k)|$ attains its maximum is found. The discrete set at the $(k+1)$st iteration is now given by $X^{k+1} = X^k \cup \mathbf{x}^k$. We will assume that initially the matrix A of the discrete problem has rank n, so that all such subsequent matrices also have rank n, and most of the methods described in Chapter 2 are available to solve these discrete subproblems. In particular, if linear programming methods are used, then the optimal basis matrix at the kth step may be used as the initial basis matrix at the $(k+1)$st step, since in the dual formulation the addition of an extra point to the set X^k corresponds to the addition of two extra variables, and the feasibility of the current solution is not affected. The process can thus be carried out efficiently; it remains to show that it is convergent.

Lemma 3.2 The sequence $\{\mathbf{a}^k\}$ generated by the first algorithm of Remes is bounded.

Proof Let $\mathbf{z}_1, \mathbf{z}_2, \ldots, \mathbf{z}_n$ be points of X^1 (and thus of all X^k) for which the corresponding $n \times n$ matrix B with (i, j) element $\phi_j(\mathbf{z}_i)$ has rank n. Now

$$f(\mathbf{z}_i) - \sum_{j=1}^{n} a_j{}^k \phi_j(\mathbf{z}_i) = r(\mathbf{z}_i, \mathbf{a}^k) \qquad i = 1, 2, \ldots, n,$$

and so

$$\mathbf{a}^k = \mathbf{B}^{-1}\mathbf{d}^k$$

where

$$|d_i{}^k| = |f(\mathbf{z}_i) - r(\mathbf{z}_i, \mathbf{a}^k)|$$
$$\leqslant 2\|f(\mathbf{x})\|$$

proving the result. $\qquad\qquad\qquad\qquad\qquad\qquad\qquad\qquad\qquad\qquad\qquad\qquad\qquad$ □

Theorem 3.8 Let $M(\mathbf{a}^k) = \max_{\mathbf{x} \in X^k} |r(\mathbf{x}, \mathbf{a}^k)|$. Then

$$M(\mathbf{a}^k) \to \min_{\mathbf{a}} \|r(\mathbf{x}, \mathbf{a})\| = M \qquad \text{as } k \to \infty.$$

Proof The sequence $\{M(\mathbf{a}^k)\}$ is obviously increasing, bounded above by M, and so tends to a limit. Assume that $M(\mathbf{a}^k) \to M - \varepsilon$, as $k \to \infty$, with $\varepsilon > 0$. Now for any $\mathbf{a}, \mathbf{d} \in R^n$, $\mathbf{x} \in X$,

$$|r(\mathbf{x}, \mathbf{a}) - r(\mathbf{x}, \mathbf{d})| = \left| \sum_{j=1}^{n} (a_j - d_j)\phi_j(\mathbf{x}) \right| \leqslant K\|\mathbf{a} - \mathbf{d}\|_2$$

where $K = n \max_{1 \leqslant j \leqslant n} \max_{\mathbf{x} \in X} |\phi_j(\mathbf{x})|$. Thus

$$|r(\mathbf{x}, \mathbf{a})| \leqslant |r(\mathbf{x}, \mathbf{d})| + K\|\mathbf{a} - \mathbf{d}\|_2.$$

Now the sequence $\{\mathbf{a}^k\}$ is bounded, by Lemma 3.2, and thus has a convergent subsequence (which we do not rename). Let $\mathbf{a}^*$ be a limit point. Then, for any $\delta > 0$, we can find k so that

$$\|\mathbf{a}^* - \mathbf{a}^k\|_2 < \delta$$

and $i > k$ so that

$$\|\mathbf{a}^* - \mathbf{a}^i\|_2 < \delta.$$

Thus

$$\|\mathbf{a}^i - \mathbf{a}^k\|_2 < 2\delta$$

and

$$
\begin{aligned}
M \leqslant \|r(\mathbf{x}, \mathbf{a}^*)\| &\leqslant \|r(\mathbf{x}, \mathbf{a}^k)\| + K\delta \\
&= |r(\mathbf{x}^k, \mathbf{a}^k)| + K\delta \\
&\leqslant |r(\mathbf{x}^k, \mathbf{a}^i)| + 3K\delta \\
&\leqslant M(\mathbf{a}^i) + 3K\delta \\
&\leqslant M - \varepsilon + 3K\delta.
\end{aligned}
$$

If $3K\delta < \varepsilon$, this gives a contradiction, and concludes the proof. $\square$

Corollary The limit points of the sequence $\{\mathbf{a}^k\}$ solve (3.1).

Example $X = [0, 1]$, $n = 1$, $\phi_1 = 1 + x$, $f(x) = x^2$.

Let $X^1 = \{0, 1\}$. We have $M(a_1{}^1) = \frac{1}{3}$, $a_1{}^1 = \frac{1}{3}$. Thus

$$r(x, a_1{}^1) = x^2 - \tfrac{1}{3}(1 + x)$$

$$Dr(x, a_1{}^1) = 2x - \tfrac{1}{3} = 0 \text{ at } x = \tfrac{1}{6},$$

when

$$r(\tfrac{1}{6}, a_1{}^1) = -13/36.$$

Thus $X^2 = \{0, \tfrac{1}{6}, 1\}$. We have $M(a_1{}^2) = 20/57 \doteq 0.350877$, $a_1{}^2 = 37/114$.
Thus $r(x, a_1{}^2) = x^2 - 37(1 + x)/114$,

$$Dr(x, a_1{}^2) = 0 \text{ at } x = 37/228.$$

when

$$\|r(x, a_1{}^2)\| \doteq 0.350896.$$

Thus $X^3 = \{0, 37/228, \tfrac{1}{6}, 1\} = \{0, 0.164, 0.167, 1\}$. The process may be continued if greater accuracy is required.

The first algorithm of Remes as described above has the advantage of being generally applicable to (3.1). However, there are situations when the method converges extremely slowly, and this is particularly so when it is applied to problems for which $\overline{E}(\mathbf{a})$ contains fewer than $(n + 1)$ points at the solution. The difficulty here is that at each step of the method, a discrete problem is being solved, and because of the rank assumption, there always exists a solution to such a problem with $(n + 1)$ points at which the maximum value is attained. Further, if linear programming is being used to solve the discrete subproblems, such a solution is always picked out in the event of non-uniqueness. Therefore the situation is that we have a sequence of solutions with $(n + 1)$ points of maximum deviation (or extrema) which we are trying to make converge to the solution of a continuous problem with at most n such points. This is in fact achieved by the coalescing of two (or perhaps more) of the extrema of the discrete problems. The optimal dual basis matrices are tending to a singular matrix, and so from a numerical point of view the subproblems become increasingly ill-conditioned.

This difficulty is particularly apparent when solving problems where X is of dimension greater than one, for $\overline{E}$ commonly contains fewer than $(n + 1)$ points at solutions. It is less common when X is an interval of the real line, although not sufficiently rare to be dismissed even in this case. Fortunately, reasonably good results can often be obtained before problems of a too seriously ill-conditioned nature have to be solved. The main drawback is then the slow convergence which is a consequence of the nature of the method as it applies in this case (see, for example, Watson, 1975; Andreassen and Watson 1976).

Example $X = [0, 2], f(x) = x^2, n = 2, \phi_1 = x, \phi_2 = e^x$.
Starting with $X^1 = \{0, 1, 2\}$, the progress of the method is summarized in Table 1. The points $x_1, x_2,$ and x_3 are the extrema of the discrete subproblem solutions.

Some improvement in the convergence rate can be obtained by modifying the method so as to define X^{k+1} as $X^k \cup \{x_i^k\}$, where $\{x_i^k\}$ is the set of *all* local maxima of $|r(x, a^k)|$ in X which are greater than $M(a^k)$. Since any method for finding the global maximum will have to investigate local maxima as part of the process, we may expect such points to be available anyway, at least approximately. Details of a procedure for their determination which may be applied in two dimensions is given by Watson (1975). Included is a search procedure which involves searching out from each of the current (discrete problem) extrema in turn in order to locate the neighbouring local maxima. Accurate values can then be obtained, for example by fitting a quadratic polynomial to the error curve and taking its maximum, or by the use of Newton's method. This part of the process is supplemented by a grid search on the edges of the region, and over parts of the interior not covered in any other way. The problem of ensuring that all the required points are located to reasonable accuracy in an efficient manner is often the most difficult part of the method to execute from a computational point of view.

If it is assumed that $\{a^k\} \to a^*$, where a^* solves (3.1), then ultimately only those points of X which are close to those in $\overline{E}(a^*)$ are really relevant. It is tempting, therefore, to *discard* some of the points as the computation proceeds. In particular, since the discrete solution at each stage is defined by just $(n + 1)$ points, this suggests that an exchange procedure in which $(n + 1)$ points are replaced by

Table 1

k	1	2	3	10	16	17	18
x_1	0	0	0.2347	0.4055	0.406356	0.406356	0.406356
x_2	1	0.4789	0.4789	0.4074	0.406395	0.406382	0.406378
x_3	2	2	2	2	2	2	2
$M(\mathbf{a}^k)$	0.4038	0.5224	0.5313	0.5382448	0.53824532	0.53824532	0.53824532
$a_1{}^k$	0.3061	−0.1911	0.0674	0.1844	0.18423193	0.18421687	0.18421184
$a_2{}^k$	0.4038	0.5224	0.4512	0.4186	0.41863139	0.41863547	0.41863683

74

another $(n+1)$ points would be appropriate. The amount of work involved in solving the linear subproblems would be considerably reduced, as would the amount of computer storage space. From a theoretical point of view, however, the convergence proof of Theorem 3.8 is no longer valid when X^k is not a subset of X^{k+1}. Thus, in general, such a procedure is not satisfactory. For the particular case when the functions $\{\phi_i(x)\}$ form a Chebyshev set on X, however, this type of approach can be made into an efficient and safe algorithm, again originally due to Remes and known as the second algorithm of Remes or multiple exchange algorithm. We need only consider the interval $[a, b]$, and we will make use of the following result which is the discrete analogue of Theorem 3.3.

Theorem 3.9 Let $X = [a, b]$ and let Y be a subset of $m > n$ points $\{x_i\}$ such that $a \leqslant x_1 < x_2 < \ldots < x_m \leqslant b$. Then if $\phi_1(x), \ldots, \phi_n(x)$ is a Chebyshev set on $[a, b]$, $\mathbf{a} \in R^n$ minimizes $\max_{x \in Y} |r(x, \mathbf{a})|$ if and only if there exists $(n+1)$ points in Y at which $\max_{x \in Y} |r(x, \mathbf{a})|$ is attained with alternating sign of $r(x, \mathbf{a})$ as we move through these points from left to right.

Proof 1 The Chebyshev set assumption means that the problem is one in R^m with a matrix A which satisfies the Haar condition. Thus in Theorem 2.1, λ has $(n+1)$ nonzero components, which, using Lemma 3.1 and Cramer's rule, alternate in sign if their ordering is that of the associated points in Y.

Proof 2 A direct proof similar to that for Theorem 3.3 may be applied. The necessity part is simplified considerably, in that intervals I_j need not be defined, but the points at which the maximum error on Y is attained dealt with directly. The details are left as an exercise. $\qquad\square$

Corollary If $m = n+1$, then the solution is obtained by solving the linear system of equations

$$f(x_i) - \sum_{j=1}^{n} a_j \phi_j(x_i) = (-1)^i h \qquad i = 1, 2, \ldots, n+1, \qquad (3.10)$$

for $\mathbf{a}$ and h.

Proof We require the nonsingularity of the $(n+1) \times (n+1)$ matrix

$$\begin{bmatrix} \phi_1(x_1) & \ldots & \phi_n(x_1) & -1 \\ \phi_1(x_2) & \ldots & \phi_n(x_2) & 1 \\ \vdots & & \vdots & \\ \phi_1(x_{n+1}) & \ldots & \phi_n(x_{n+1}) & (-1)^{n+1} \end{bmatrix},$$

which follows by expanding the determinant by the last column and using Lemma 3.1. Alternatively, the assumption of singularity is easily seen to lead to a contradiction (see also exercise 8 of Chapter 2). Notice that if $h < 0$, then the minimum value of the norm on R^{n+1} is $-h$. ☐

We now describe the second algorithm of Remes, bearing in mind that the set $\{\phi_i(x)\}$ is assumed to be a Chebyshev set on $X = [a, b]$. At the kth step, a subset X^k of X containing precisely $(n+1)$ distinct points is defined. The discrete problem on these points is solved (i.e. the system (3.10) is solved). The error curve $r(x, \mathbf{a}^k)$, where $\mathbf{a}^k$ is the solution obtained to the problem on X^k, is examined, and $(n+1)$ local maxima of $|r(x, \mathbf{a}^k)|$ obtained in X, with values not exceeded on X^k, such that (a) the values of $r(x, \mathbf{a}^k)$ have alternate sign as we move along these points from left to right, and (b) a (global) maximum point of $|r(x, \mathbf{a}^k)|$ is included. This set of points forms the set X^{k+1}.

It remains to show that such a set can always be obtained. Starting from any point $x_i^k \in X^k$, it is clear that by searching out from this point, a local maximum of $|r(x, \mathbf{a}^k)|$ with sign the same as that of $r(x_i^k, \mathbf{a}^k)$ must, by continuity, always be obtained. (This may be just the point x_i^k, of course). Thus we can certainly locate $(n+1)$ points at which the signs of $r(x, \mathbf{a}^k)$ alternate. If this set includes a global maximum point of $|r(x, \mathbf{a}^k)|$, then we have the set X^{k+1}. Otherwise such a point must be included, and this may be done by exchanging it for an adjacent local maximum having the same sign. If no global maximum point lies between two of the local maxima obtained, then either the leftmost or rightmost of these points must be deleted. In any event, it is clearly always possible to achieve the desired objective, and define X^{k+1}. A typical situation is illustrated in Figure 9 (case $n = 4$). The points y_i^k are those which would be obtained initially in the manner described. Since $x = a$ gives the global maximum of $|r(x, \mathbf{a}^k)|$, then it replaces y_1^k in the set X^{k+1}.

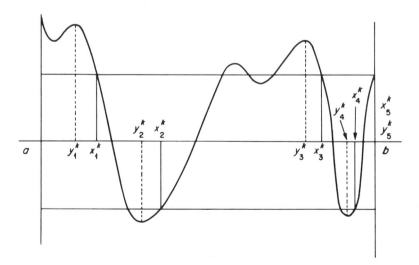

Figure 9

As already noted, the convergence proof of Theorem 3.8 is no longer valid in this case, because of the fact that previous points are not retained. However convergence still holds, as the next result shows.

Theorem 3.10 Let $M(\mathbf{a}^k) = \max_{x \in X^k} |r(x, \mathbf{a}^k)|$. Then

$$M(\mathbf{a}^k) \to \min_{\mathbf{a}} \| r(x, \mathbf{a}) \| \equiv M \qquad \text{as } k \to \infty.$$

Proof

$$f(x_i{}^k) - \sum_{j=1}^n a_j{}^k \phi_j(x_i{}^k) = \alpha(-1)^i M(\mathbf{a}^k) \qquad i = 1, 2, \ldots, n+1, \quad (3.11)$$

where $\alpha = \pm 1$. Now

$$f(x_i{}^{k+1}) - \sum_{j=1}^n a_j{}^k \phi_j(x_i{}^{k+1}) = \beta(-1)^i \varepsilon_i{}^k \qquad i = 1, 2, \ldots, n+1, \quad (3.12)$$

where $\beta = \pm 1, \varepsilon_i{}^k \geqslant M(\mathbf{a}^k), i = 1, 2, \ldots, n+1$ with strict inequality holding for at least one component, unless $\mathbf{a}^k$ solves the original problem. Further, for some $\gamma = \pm 1$.

$$f(x_i{}^{k+1}) - \sum_{j=1}^n a_j{}^{k+1} \phi_j(x_i{}^{k+1}) = \gamma(-1)^i M(\mathbf{a}^{k+1}) \qquad i = 1, 2, \ldots, n+1.$$

$$(3.13)$$

For the set X^{k+1}, let λ^{k+1} satisfy the conditions of Theorem 2.1 and be normalized so that $\sum_{i=1}^{n+1} |\lambda_i{}^{k+1}| = 1$. Then, using (3.13) and (3.12),

$$M(\mathbf{a}^{k+1}) = \sum_{i=1}^{n+1} |\lambda_i{}^{k+1}| \varepsilon_i{}^k.$$

Thus

$$M(\mathbf{a}^{k+1}) - M(\mathbf{a}^k) = \sum_{i=1}^{n+1} |\lambda_i{}^{k+1}| (\varepsilon_i{}^k - M(\mathbf{a}^k))$$
$$> 0.$$

It follows that $\{M(\mathbf{a}^k)\}$ is an increasing sequence, bounded above, and so convergent. Now by the Chebyshev set assumption, we have

$$|\lambda_i{}^{k+1}| > 0 \qquad i = 1, 2, \ldots, n+1.$$

In fact there exists C independent of i, k such that

$$|\lambda_i{}^{k+1}| \geqslant C > 0 \qquad \text{all } i, \text{ all } k. \qquad (3.14)$$

For if such a C does not exist, then the sets X^k cannot contain points which remain strictly separated, and so there must exist a limiting set of $(n+1)$ points in which 2 or more points are identical. For this set, the system (3.10) must have a solution with $h = 0$, contradicting the fact that successive values of $M(\mathbf{a}^k)$ are monotonically increasing. Thus (3.14) holds and so

$$\sum_{i=1}^{n+1} |\lambda_i^{k+1}|(\varepsilon_i^k - M(\mathbf{a}^k)) \geqslant C(\|r(x, \mathbf{a}^k)\| - M(\mathbf{a}^k)).$$

Therefore

$$\|r(x, \mathbf{a}^k)\| - M(\mathbf{a}^k) \to 0 \quad \text{as} \quad k \to \infty.$$

Now

$$0 \leqslant M - M(\mathbf{a}^k) \leqslant \|r(x, \mathbf{a}^k)\| - M(\mathbf{a}^k)$$

and the result of the thorem follows. $\qquad\square$

Theorem 3.11

$$\lim_{k \to \infty} \mathbf{a}^k = \mathbf{a}^* \qquad \text{where } \mathbf{a}^* \text{ solves (3.1)}.$$

Proof By Theorem 3.6, there exists $\gamma > 0$ such that

$$\|r(x, \mathbf{a}^k)\| \geqslant \|r(x, \mathbf{a}^*)\| + \gamma \|\mathbf{a}^k - \mathbf{a}^*\|_2.$$

From Theorem 3.10,

$$\|r(x, \mathbf{a}^k)\| - \|r(x, \mathbf{a}^*)\| \to 0 \quad \text{as} \quad k \to \infty,$$

and so the result follows. $\qquad\square$

If the functions $f(x), \phi_1(x), \ldots, \phi_n(x)$ are twice continuously differentiable, then the second algorithm of Remes has a second order convergence rate. This result is due to Veidinger (1960). We prove it for the special case when the end points of $[a, b]$ are members of the optimal alternating set; only minor changes are necessary to take account of the other possibilities. The particular form of proof is due to Meinardus (1967).

Theorem 3.12 Let $f(x), \phi_1(x), \ldots, \phi_n(x) \in C^2[a, b]$, let $\mathbf{a}$ solve (3.1) with $\overline{E}(\mathbf{a}) = \{x_1, x_2, \ldots, x_{n+1}\}$, where $x_1 = a$ and $x_{n+1} = b$, and assume that

$$D^2 r(x_i, \mathbf{a}) \neq 0 \qquad i = 2, 3, \ldots, n$$
$$Dr(x_i, \mathbf{a}) \neq 0 \qquad i = 1, n+1.$$

Then

$$\|\mathbf{a}^{k+1} - \mathbf{a}\|_2 \leqslant K \|\mathbf{a}^k - \mathbf{a}\|_2^2$$

where K is a constant.

Proof At the solution to (3.1)

$$Dr(x_i, \mathbf{a}) = 0 \qquad i = 2, 3, \ldots, n,$$

which, using the implicit function theorem, may be regarded as defining $x_2, x_3, \ldots, x_n$ as (differentiable) functions of $\mathbf{a}$. Thus the equations

$$r(x_i, \mathbf{a}) - (-1)^i h = 0, \qquad i = 1, 2, \ldots, n+1 \qquad (3.15)$$

satisfied by $|h| = \|r(x, \mathbf{a})\|$ can be thought of as a nonlinear system of equations in $\mathbf{a}$ and h. By continuity, for $\|\mathbf{a}^k - \mathbf{a}\|_2$ sufficiently small, we may uniquely define x_i^{k+1}, $i = 2, 3, \ldots, n$ by

$$Dr(x_i^{k+1}, \mathbf{a}^k) = 0 \qquad i = 2, 3, \ldots, n, \qquad (3.16)$$

and applying Newton's method to (3.15) at approximations to $\mathbf{a}$ and h given by $\mathbf{a}^k$ and h^k respectively, we have the new approximations $\mathbf{a}^{k+1}$ and h^{k+1} satisfying

$$r(x_i^{k+1}, \mathbf{a}^{k+1}) = (-1)^i h^{k+1} \qquad i = 1, 2, \ldots, n+1. \qquad (3.17)$$

Equation (3.16) and (3.17) just define the new approximations obtained by the second algorithm of Remes, for eventually we must have $x_1^{k+1} = a, x_{n+1}^{k+1} = b$. The result of the theorem follows from the convergence properties of the Newton iteration. $\square$

Example $X = [0, 1]$, $f(x) = 1/(1 + x)$, $n = 3$, $\phi_1 = 1$, $\phi_2 = x$, $\phi_3 = x^2$.

k	1	2	3
x_1^k	0.0	0.0	0.0
x_2^k	0.3	0.2016	0.2071
x_3^k	0.6	0.7149	0.7071
x_4^k	1.0	1.0	1.0
$M(\mathbf{a}^k)$	0.006 181	0.007 354	0.007 360
$\|r(x, \mathbf{a}^k)\|$	0.008 013	0.007 363	0.007 360

It is clear that 3 iterations of the method have given the minimum norm accurate to 6 decimal places. The initial approximation (or equivalently the set X^1) should obviously be chosen as near as possible to the optimal extremal set. In the absence of any information, equispaced or nearly equispaced points may be used, although for polynomial approximation a better choice exists (see Section 5.4). The decision when to terminate the process could be based on the inequality

$$\|r(x, \mathbf{a}^k)\| - M(\mathbf{a}^k) \leqslant \delta$$

being satisfied, for some given value of δ. For a modern implementation of the method, see Golub and Smith (1971).

The second algorithm of Remes may be used for the calculation of approximations to functions of two variables. Let $X = X_1 \times X_2$, let $f(x, y) \in C[X]$ be given, and let $\phi_i(x)$, $i = 1, 2, \ldots, n$ be a Chebyshev set on X_1, and $\psi_i(y)$,

$i = 1, 2, \ldots, n$ be a Chebyshev set on X_2. For given $y \in X_2$, let $\mathbf{a}(y)$ minimize $\max\limits_{x \in X_1}$

$\left| f(x, y) - \sum\limits_{i=1}^{n} a_i \phi_i(x) \right|$ and let $\mathbf{b}_j \in R^n$ minimize $\max\limits_{y \in X_2} \left| a_j(y) - \sum\limits_{i=1}^{n} (\mathbf{b}_j)_i \psi_i(y) \right|$. Then

$\sum\limits_{j=1}^{n} \sum\limits_{i=1}^{n} (\mathbf{b}_j)_i \psi_i(y) \phi_j(x)$ is called the best *product* Chebyshev approximation to $f(x, y)$ on X. For further details of this, see for example Weinstein (1969, 1971).

Exercises

38. Explain how the exchange method of Stiefel described in Chapter 2 may be applied to the problems treated in this chapter, and determine its connection with the algorithms of Remes.

39. Modify the second proof of Theorem 3.3 to prove Theorem 3.9.

40. Let P_1 and P_2 be polynomials of degree $(n+1)$ defined by the interpolation conditions

$$P_1(x_i) = f(x_i)$$
$$P_2(x_i) = (-1)^i$$

for $i = 1, 2, \ldots, n+2$ where $x_1 < x_2 < \ldots < x_{n+2}$. Prove that the polynomial of degree n which gives the best Chebyshev approximation to $f(x)$ on $x = x_i$, $i = 1, 2, \ldots, n+2$ is given by $P_1 + \alpha P_2$, where α is chosen to make the coefficient of x^{n+1} zero.

41. Prove Theorem 3.11 without explicitly using the strong uniqueness result.

42. Modify the statement and proof of Theorem 3.12 to take account of the other possible positions of the points in $\overline{E}(\mathbf{a})$.

43. Use both the first and second algorithms of Remes to find the best Chebyshev approximation to $x^{\frac{1}{2}}$ on $[0, 1]$ by a straight line. Take as initial point set $\{1/9, 1/4, 9/16\}$. (A hand calculator is all that is required here.)

44. Consider carefully the problems involved in computing the local (including the global) maximum modulus points of a continuously differentiable function on the interval $[0, 1]$. Write a program to implement the second algorithm of Remes for approximation on this interval, and find the best polynomial approximations of given degree to the following functions

   (i) $n = 2$ $f(x) = x^4$
  (ii) $n = 3$ $f(x) = \sin x$
 (iii) $n = 4$ $f(x) = \cos x$.

45. Use the second algorithm of Remes to obtain an approximate solution, in

the form of a polynomial, to the differential equation

$$2(1 + x)Dy + y = 0 \qquad y(0) = 1, 0 \leqslant x \leqslant 1.$$

(Ignore the initial condition, and scale the solution at the end.)

3.6 Some other approaches

In the absence of Chebyshev sets, the first algorithm of Remes is, as we have seen, available for the computation of best approximations. The method is, however, slow in cases where $\overline{E}(\mathbf{a})$, where $\mathbf{a}$ solves (3.1), contains fewer than $(n + 1)$ points, for reasons which have already been explained. One way in which a fast convergence rate can sometimes be recovered is through the direct satisfaction of the conditions of Theorem 3.1. The method which is now described requires that $f(x)$ and $\phi_j(x), j = 1, 2, \ldots, n$, be continuously differentiable.

Consider the nonlinear system of equations

$$f(\mathbf{x}_i) - \sum_{j=1}^{n} a_j \phi_j(\mathbf{x}_i) = \hat{\theta}_i h \qquad i = 1, 2, \ldots, t,$$

$$\sum_{i=1}^{t} \lambda_i \phi_j(\mathbf{x}_i) = 0 \qquad j = 1, 2, \ldots, n, \qquad (3.18)$$

$$\sum_{i=1}^{t} \lambda_i \hat{\theta}_i = 1,$$

where $\hat{\theta}_i$ are given numbers satisfying $|\hat{\theta}_i| = 1, i = 1, 2, \ldots, t$, for unknowns $\mathbf{x}_i$, $i = 1, 2, \ldots, t, \lambda_i, i = 1, 2, \ldots, t, a_j, j = 1, 2, \ldots, n$, and h. Now if a solution can be found satisfying

$$\text{sign}\,(r(\mathbf{x}_i, \mathbf{a})) = \hat{\theta}_i \qquad i = 1, 2, \ldots, t$$
$$\lambda_i \hat{\theta}_i \geqslant 0 \qquad i = 1, 2, \ldots, t,$$

and $\|r(\mathbf{x}, \mathbf{a})\| = h$, then the conditions of Theorem 3.1 are satisfied, and $\mathbf{a}$ solves (3.1). The system (3.18) is of course underdetermined, since it represents $(n + t + 1)$ equations in $Nt + n + t + 1$ unknowns. However, if $\mathbf{x}_i$ is an *internal* point of X, then the N partial derivatives of $r(\mathbf{x}, \mathbf{a})$ with respect to the components of $\mathbf{x}$ must vanish there. If $\mathbf{x}_i$ lies on an *edge* of X, then again N pieces of information will be available in the form of vanishing partial derivatives and/or knowledge of certain of the components of $\mathbf{x}_i$. Thus we can supplement the system (3.18) by Nt additional equations, so that we have the same number of equations as unknowns. If $f(\mathbf{x}), \phi_j(\mathbf{x}), j = 1, 2, \ldots, n$, are *twice* continuously differentiable, then Newton's method, for example, may be used to solve the problem.

Details of a method such as this are given in Andreassen and Watson (1976), based on the use of the first algorithm of Remes to provide good initial approximations. Fast ultimate convergence requires that the Jacobian matrix of the system be nonsingular at the solution to which the method is converging.

However, this kind of assumption can be avoided by the use of more robust Newton-type methods. In practice the method appears to work well.

Example $X = [0, 2], f(x) = x^2, n = 2, \phi_1 = x, \phi_2 = e^x$.

The information in Table 1 can be used to provide starting values for Newton's method applied to the appropriate nonlinear system. If the starting values are taken from the approximation at iteration number 3, Newton's method provides the following values of a_1, a_2, and h.

a_1	0.0674	0.1799	0.184230	0.184232564
a_2	0.4512	0.4200	0.4186321	0.418631218
h	0.5313	0.5369	0.538244	0.538245318

The final values are correct to the number of figures shown.

A method for (3.1) based on the Polya algorithm has been implemented by Fletcher, Grant, and Hebden (1974a) (see also Karlovitz, 1978). This is the analogue of the same method described in the previous chapter, where a sequence of L_p problems is solved for increasing values of p. In the continuous case, convergence is not guaranteed, but appears to occur for reasonable problems (Descloux, 1963, gives an example of nonconvergence involving a function with an infinite number of discontinuous derivatives). The main drawback of the method is that convergence is slow.

Descent methods analogous to those described in Chapter 2 are also available in the continuous case. They are not so attractive here because the finite nature of the computation in the discrete case is lost, although some success in using such a method is claimed by Scott and Thorp (1972). A descent direction $\mathbf{c}$ is obtained by satisfying the system of inequalities (3.3), and a step taken in this direction so that the value of the norm is reduced. The nature of the problem in this case is such that no reliable method for computing such a step length suggests itself, although use may be made of the bounds for γ obtained in the proof of necessity in Theorem 3.1.

Some comparisons undertaken by Andreassen and Watson (1976) suggest that methods of Remes type should be used for the problem (3.1), particularly if they are supplemented where necessary by a method such as the one described at the start of this section. In their present form the other methods do not appear to be competitive.

3.7 Muntz–Jackson theorems

For a particular set of functions $\phi_1(\mathbf{x}), \ldots, \phi_n(\mathbf{x})$ of interest is the general problem of obtaining *a priori* information on how *well* a given function $f(\mathbf{x})$ can be approximated. Such information is, for example, useful in deciding whether or not the particular set $\{\phi_i(\mathbf{x})\}$ is indeed appropriate. Defining $E_n(f, \phi)$ by

$$E_n(f, \phi) = \min_{\mathbf{a} \in R^n} \left\| f(\mathbf{x}) - \sum_{i=1}^{n} a_i \phi_i(\mathbf{x}) \right\|, \qquad (3.19)$$

we would therefore like to obtain best possible inequalities of the form

$$E_n(f, \phi) \leqslant g(f, n). \qquad (3.20)$$

If a linearly independent *sequence* $\{\phi_i(\mathbf{x})\}$ can be defined, then the behaviour of $E_n(f, \phi)$ as $n \to \infty$ may be deduced from that of $g(f, n)$. The simplest example of such a sequence is that given in $X = [a, b]$ by $\phi_i(x) = x^{i-1}, i = 1, 2, \ldots$ when it is known that $E_n(f, \phi) \to 0$ as $n \to \infty$ as a consequence of the Weierstrass Theorem. Another way of stating this is that $\{1, x, x^2, \ldots\}$ is fundamental in $C[a, b]$. The provision of inequalities of the type (3.20) in this case was considered in detail by Jackson in 1911, who gave a number of different versions depending on the particular assumptions made about $f(x)$. Subsequent results of this form which have been obtained are referred to as Jackson-type theorems.

The sequence $\{1, x, x^2, \ldots\}$ is actually a special case of a more general sequence of powers of x which also has the property that $E_n(f, \phi) \to 0$ as $n \to \infty$. The following theorem, which we do not prove, was given by Muntz in 1914.

Theorem 3.13 The sequence $\{1, x^{\lambda_1}, x^{\lambda_2}, \ldots\}$, where $\{\lambda_k\}$ is an increasing sequence of positive numbers, is fundamental in $C[a, b]$ if and only if

$$\sum_{k=1}^{\infty} \frac{1}{\lambda_k}$$

diverges.

Theorems giving inequalities of the form (3.20) for approximation by linear combinations of elements of this sequence are known as Muntz–Jackson theorems. A large number of such results have appeared in recent years, with numerous generalizations from the L_∞ to the L_p norms, and, in other directions, to rational function approximations formed by the quotient of linear combinations of elements of the above sequence. Some of these results will be mentioned elsewhere in this book, but meantime we will merely give an indication of the general form, and also the source, of some results which are now available for the L_∞ norm.

To maintain compatibility with the majority of published results for this problem, we will define $E_n(f)$ as

$$E_n(f) = \min_{\mathbf{a} \in R^{n+1}} \left\| f(x) - \sum_{i=1}^{n+1} a_i x^{i-1} \right\|, \qquad (3.21)$$

so that the approximating function is a $\cdot$*polynomial of degree n*, and we will also define $E_n(f, \Lambda)$ by

$$E_n(f, \Lambda) = \min_{\mathbf{a} \in R^{n+1}} \left\| f(x) - a_1 - \sum_{i=2}^{n+1} a_i x^{\lambda_{i-1}} \right\|, \qquad (3.22)$$

where Λ is used to denote the sequence $\{\lambda_1, \lambda_2, \ldots, \lambda_n\}$. Thus

$$E_n(f) = E_n(f, \Lambda)$$

if $\Lambda = \{1, 2, \ldots, n\}$. If $\omega_k(\delta)$ denotes the modulus of continuity of $D^k f(x)$, defined by

$$\omega_k(\delta) = \sup_{|x-y| \leqslant \delta} |D^k f(x) - D^k f(y)|,$$

then if $f(x) \in C^k[-1, 1]$ the classical results of Jackson can be given in the form

$$E_n(f) \leqslant A_k n^{-k} \omega_k\left(\frac{1}{n}\right) \tag{3.23}$$

where $k < n$, and A_k is a constant, independent of f and n. In particular, for the case $k = 0$, it is known that we can take

$$A_0 = 1 + \frac{\pi^2}{2}$$

(see Meinardus, 1967, for a proof of this) although this constant is not the best possible. Other forms of these theorems are given by Cheney (1966). For example, it is proved there that if $f(x) \in C^k[-1, 1]$ and $n \geqslant k$,

$$E_n(f) \leqslant (\pi/2)^k \|D^k f\| / ((n+1)(n) \ldots (n-k+2)). \tag{3.24}$$

When Λ is a sequence satisfying the conditions of Theorem 3.13, and also the condition that $\lambda_k \geqslant 2k$, then Bak and Newman (1972) obtain the result

$$E_n(f, \Lambda) \leqslant A\omega_0(\delta),$$

where A is constant. Leviatan (1974) gives bounds for $E_n(f, \Lambda)$ when $f(x) \in C^k[0, 1]$ in terms of the modulus of continuity $\omega_k(\delta)$, and further generalizations are given by Golitschek (1976). The results become rather cumbersome, and we merely refer to these papers.

A number of Jackson-type theorems are also available when $f(x)$ is a *particular* function. For example, Bernstein in 1912 considered $f(x) = |x|$, and obtained the inequality

$$E_n(f) \leqslant C/n$$

where C is a constant. The case $f(x) = x^\alpha$, where $\alpha > 0$ is considered by Elosser (1978), who generalizes some earlier results for this function. For $1 < \alpha < n$, he obtains the inequality

$$E_n(x^\alpha) < \frac{1}{2(n - [\alpha - 1]) + 1}$$

where $[\alpha - 1]$ is the greatest integer $\leqslant \alpha - 1$, and for $\alpha \in [n, n+1]$,

$$E_n(x^\alpha) \leqslant \frac{1}{2^{2n+1}}.$$

In addition to upper bounds for $E_n(f)$ and $E_n(f, \Lambda)$, a number of authors have also considered *lower bounds* for these quantities (see also exercises 15 and 16). Such theorems can give information on how *slowly* $E_n(f)$, for example, tends to zero; where arbitrary slowness is possible, the corresponding results are referred to as '*lethargy theorems*'. Indeed, $E_n(f)$ can converge to zero very slowly for some continuous functions: Bernstein (1938) shows that if $\{\varepsilon_n\}$ is any positive sequence converging to zero, then there exists $f(x) \in C[-1, 1]$ such that $E_n(f) = \varepsilon_n$, for all n.

Exercises

46.   Let $p_n(x)$ be the polynomial of degree n which interpolates to $f(x) \in C[a, b]$ at the points x_i, $i = 1, 2, \ldots, n+1$ in $[a, b]$ and define $e_n(f)$ by

$$e_n(f) = \| f(x) - p_n(x) \|.$$

Prove that

$$e_n(f) \leqslant E_n(f) \left(1 + \max_{a \leqslant x \leqslant b} \sum_{i=1}^{n+1} |l_i(x)| \right),$$

where $\{l_i(x)\}$ are the Lagrange coefficients defined in Chapter 1. (Powell, 1967).

47.   Let $f(x) \in C^{(n+1)}[-1, 1]$. Prove that there exists $\xi \in (-1, 1)$ such that

$$E_n(f) = \frac{2^{-n} |D^{n+1} f(\xi)|}{(n+1)!}$$

(Meinardus, 1967; see also Phillips, 1968).

4

The L_p solution of an overdetermined system of linear equations

4.1 Introduction

We return now to approximation in the space R^m, normed with the L_p norm

$$\|\mathbf{r}\| = \left(\sum_{i=1}^{m} |r_i|^p \right)^{1/p}$$

where $1 < p < \infty$. The most important example of this occurs when $p = 2$ when we have the least squares norm, and this will be given special treatment. However, in many respects it is possible to treat the problem in a general fashion without choosing specific values of p from the given range. In particular, as we have already seen, the space defined above is strictly convex for $1 < p < \infty$ and so a special study of uniqueness does not arise. In addition, if $\|\mathbf{r}\| \neq 0$ then $\|\mathbf{r}\|$ is a differentiable function of the components of $\mathbf{r}$, and so a particularly convenient form of characterization can be easily obtained.

The basic approximation problem which we treat can be stated:

$$\text{find } \mathbf{a} \in R^n \text{ to minimize } \|\mathbf{r}(\mathbf{a})\| \tag{4.1}$$

where

$$\mathbf{r}(\mathbf{a}) = \mathbf{b} - A\mathbf{a}, \tag{4.2}$$

with A a given $m \times n$ matrix, and $\mathbf{b} \in R^m$ also given. For particular $\mathbf{a} \in R^n$, let D_1 denote the $m \times m$ diagonal matrix with (i, i) element $|r_i|^{p-1}, i = 1, 2, \ldots, n$ and let $\boldsymbol{\theta} \in R^m$ be such that

$$\theta_i = \text{sign}(r_i) \qquad i = 1, 2, \ldots, m.$$

Theorem 4.1 A vector $\mathbf{a} \in R^n$ solves (4.1) if and only if

$$A^T D_1 \boldsymbol{\theta} = \mathbf{0}. \tag{4.3}$$

(If $\|\mathbf{r}\| = 0$ the result is trivial, so assume not.)

Proof 1 In the notation of Theorem 1.7, $\mathbf{v} \in V(\mathbf{r})$ if and only if it is the unique vector in R^m with ith component

$$v_i = \theta_i |r_i|^{p-1} \|\mathbf{r}\|^{1-p} \qquad i = 1, 2, \ldots, m.$$

The conditions of Theorem 1.7 immediately give (4.3).

Proof 2 The point minimizing $\|\mathbf{r}\|$ also minimizes $\|\mathbf{r}\|^p$, which is a differentiable function of $\mathbf{r}$, and therefore of $\mathbf{a}$. Thus if $\mathbf{a}$ is a minimum it must satisfy

$$p \sum_{i=1}^{m} |r_i|^{p-1} \theta_i (-\alpha_i) = \mathbf{0}$$

which gives (4.3). If $\mathbf{a}$ satisfies (4.3), then since $\|\mathbf{r}\|^p$ is a convex function of $\mathbf{a}$, exercise 15 of Chapter 1 shows that $\mathbf{a}$ solves (4.1), and this concludes the proof. $\qquad\square$

A necessary and sufficient condition for a unique $\mathbf{a}$ solving (4.1) is that A have rank n.

4.2 The case $p = 2$

The problem (4.1) for $p = 2$ is the well-known linear least squares problem, and this frequently arises in the context of discrete data analysis. Statistically, the use of the least squares norm is justified when the expected values of r_i are zero, and these values represent errors which are independent and normally distributed with constant variance σ^2. The most likely explanation of the data is then obtained by maximizing what is essentially the product of the probability density functions

$$f(r_i) = \frac{1}{\sqrt{(2\pi)}\sigma} \exp\left(-\frac{r_i^2}{2\sigma^2}\right) \qquad i = 1, 2, \ldots, m;$$

clearly this is achieved by minimizing $\|\mathbf{r}\|_2$. The system of equations (4.3) is in this case just a linear system of n equations in the components of $\mathbf{a}$ called the *normal equations*.

The 'method of least squares' was first published by Legendre in 1801. It was also mentioned in published work by Gauss in 1809, who claimed to have been using the method since 1795. Some discussion then resulted about who was the true originator of the method, although nowadays any argument is generally accepted as having been resolved in favour of Gauss.

The system of equations (4.3) can be written

$$A^T\mathbf{r} = \mathbf{0}$$

giving

$$A^TA\mathbf{a} = A^T\mathbf{b}, \tag{4.4}$$

the familiar form of the normal equations. If A has rank n, then A^TA is symmetric positive definite (exercise 1), and the solution to (4.4) can be obtained using the Cholesky decomposition of the matrix A^TA. Thus, in theory, the solution of the problem is straightforward. It has, however, been appreciated for some time that the matrix A^TA can be ill-conditioned with respect to inversion. The effect of this is to amplify any rounding errors which arise in the numerical solution, and so prevent an accurate solution of the system. In view of this, the direct solution of (4.4) can often be an extremely *bad* way of trying to solve the least squares problem.

Example

$$A = \begin{bmatrix} 1 & 1 & 1 \\ \varepsilon_1 & 0 & 0 \\ 0 & \varepsilon_2 & 0 \\ 0 & 0 & \varepsilon_3 \end{bmatrix}$$

If $|\varepsilon_i| \leqslant 10^{-6}$, $i = 1, 2, 3$, and the computation is assumed to be performed on a computer whose word length is such that no more than 10 decimal places are retained, then

$$A^TA = \begin{bmatrix} 1+\varepsilon_1{}^2 & 1 & 1 \\ 1 & 1+\varepsilon_2{}^2 & 1 \\ 1 & 1 & 1+\varepsilon_3{}^2 \end{bmatrix}$$

will be stored in the computer as the singular matrix each of whose elements is unity.

This (extreme) example is intended to illustrate the essential change which can take place in the *condition* of the least squares problem by the formation of the matrix A^TA. To be more precise, we introduce the measure of the condition known as the *condition number*. For a square matrix C, this may be given by the expression

$$K(C) = \|C\| \, \|C^{-1}\|$$

where we define the matrix norm by

$$\|C\| = \max_{\|\mathbf{x}\| = 1} \|C\mathbf{x}\|$$
$$= (\text{maximum eigenvalue of } C^TC)^{\frac{1}{2}}$$

(exercise 3), which is often referred to as the *spectral norm* of C. The significance of this particular measure is illustrated by the result (exercise 4) that if the nonsingular system of linear equations

$$Cx = c \qquad c \neq 0,$$

has the exact solution x^*, then for any $x \in R^n$,

$$\frac{\|x - x^*\|}{\|x^*\|} \leq K(C)\frac{\|Cx - c\|}{\|c\|}. \qquad (4.5)$$

Thus even if $\|Cx - c\|$ is close to zero, if $K(C)$ is large x may be considerably different from x^*.

To extend the definition of condition number to a nonsquare matrix C, we write

$$K(C) = \max_{\|x\| = 1} \|Cx\| \Big/ \min_{\|x\| = 1} \|Cx\|$$
$$= \sigma_1/\sigma_n,$$

where $\sigma_1{}^2$ and $\sigma_n{}^2$ are respectively the greatest and least eigenvalues of C^TC. When C is square, this expression reduces to that defined before. According to this definition

$$K(A^TA) = K(A)^2;$$

thus if $K(A)$ is large, the inherent difficulty in the solution of (4.4) is clear.

An alternative procedure for the least squares problem which avoids the formation of A^TA is due in its original form to Golub (1965). We will use the following lemma, the proof of which is left as an exercise.

Lemma 4.1 Let $x \in R^k$, $y \in R^k$ satisfy $\|x\| = \|y\|$, $x \neq y$, and let $Q = I - 2ww^T$, where $w = \lambda(x - y)$ with λ chosen so that $\|w\| = 1$. Then Q is a $k \times k$ orthogonal matrix satisfying $Qx = y$.

The matrix Q is usually referred to as an elementary Householder matrix, and the use of such matrices in solving (4.1) is in achieving the following factorization.

Theorem 4.2 There exists an $m \times m$ orthogonal matrix W such that

$$A = W\begin{bmatrix} U \\ 0 \end{bmatrix}$$

where U is an $n \times n$ upper triangular matrix, and 0 represents an $(m - n) \times n$ block of zero elements.

Proof We will give a constructive proof of this result. In Lemma 4.1, let $k = m$, let x be the first column of A, and if $x \neq 0$, let W_1 be the elementary Householder matrix satisfying

$$W_1 x = -\sigma\|x\|e_1$$

where $\sigma = +1$ if $x_1 \geq 0$ and -1 otherwise. Then $A_1 = W_1 A$ is a matrix whose first column has zeros below the (1, 1) element. (If $\mathbf{x} = \mathbf{0}$, then $A_1 = A$.) Now in Lemma 4.1, let $k = m - 1$, let $\mathbf{x} \in R^k$ consist of components 2 to m of the second column of A_1, and if $\mathbf{x} \neq \mathbf{0}$, let Q_2 be the elementary Householder matrix satisfying

$$Q_2 \mathbf{x} = -\sigma \| \mathbf{x} \| \mathbf{e}_1$$

with σ defined as before. Then

$$W_2 = \begin{bmatrix} 1 & 0 \\ 0 & Q_2 \end{bmatrix}$$

is an orthogonal matrix with $A_2 = W_2 A_1$ having zeros below the main diagonal in the first two columns. If we continue in this way, after (at most) n steps we will have the required result. $\qquad\square$

Now let the rank of A be $t < n \leqslant m$, and let P be an $n \times n$ permutation matrix which permutes the columns of A so that the first t columns of AP are linearly independent (see, for example, Gourlay and Watson, 1973). Then if we apply Theorem 4.2 to AP we can write

$$\hat{W} AP = \begin{bmatrix} \hat{U} \\ 0 \end{bmatrix},$$

where $\hat{W}$ is an $m \times m$ orthogonal matrix, and $\hat{U}$ is an $n \times n$ upper triangular matrix. Since the first t columns of $\hat{W} AP$ are linearly independent, this matrix has the form $\begin{bmatrix} U_1 & V_1 \\ 0 & 0 \end{bmatrix}$, where U_1 is a nonsingular $t \times t$ upper triangular matrix. Thus

$$\hat{W} AP = \begin{bmatrix} L & 0 \\ 0 & 0 \end{bmatrix} \hat{Q}$$

applying Theorem 4.2 to $[U_1 \, V_1]^T$, where L is a nonsingular $t \times t$ lower triangular matrix, and $\hat{Q}$ is an $n \times n$ orthogonal matrix. It follows that if rank (A) $= t \leqslant n \leqslant m$, we can write

$$A = W \begin{bmatrix} T & 0 \\ 0 & 0 \end{bmatrix} Q$$

where W is an $m \times m$ orthogonal matrix, Q is an $n \times n$ orthogonal matrix and T is a $t \times t$ nonsingular triangular matrix. Then

$$\| \mathbf{r} \|^2 = (A\mathbf{a} - \mathbf{b})^T (A\mathbf{a} - \mathbf{b})$$

$$= \left(\begin{bmatrix} T & 0 \\ 0 & 0 \end{bmatrix} Q\mathbf{a} - W^T \mathbf{b} \right)^T \left(\begin{bmatrix} T & 0 \\ 0 & 0 \end{bmatrix} Q\mathbf{a} - W^T \mathbf{b} \right)$$

$$= \begin{bmatrix} T\mathbf{d}_1 - \mathbf{c}_1 \\ -\mathbf{c}_2 \end{bmatrix}^T \begin{bmatrix} T\mathbf{d}_1 - \mathbf{c}_1 \\ -\mathbf{c}_2 \end{bmatrix}$$

where $W^T\mathbf{b} = \begin{bmatrix} \mathbf{c}_1 \\ \mathbf{c}_2 \end{bmatrix}$, $Q\mathbf{a} = \mathbf{d}$, and $\mathbf{d}_1$ is the vector formed by the first t components of $\mathbf{d}$. Thus the sum of squares is minimized, with value $\mathbf{c}_2{}^T\mathbf{c}_2$, when

$$T\mathbf{d}_1 = \mathbf{c}_1,$$

which is a nonsingular triangular $t \times t$ system for $\mathbf{d}_1$. Finally,

$$\mathbf{a} = Q^T\mathbf{d},$$

where the last $(n-t)$ components of $\mathbf{d}$ are arbitrary.

It is possible (though we will not pursue this in detail) to further decompose T into the form

$$T = MSN,$$

where S is a nonsingular $t \times t$ diagonal matrix, and M and N are $t \times t$ orthogonal matrices. The diagonal elements of S are the positive singular values of A (the square roots of the eigenvalues of A^TA) and the resulting orthogonal decomposition of A is known as the *singular value decomposition*, which is particularly useful in an analysis of the influence on solutions to (4.1) of errors in the data. The use of orthogonal transformation matrices to effect these decompositions of A not only permits (4.1) to be solved in very general circumstances, but also results in algorithms which are numerically stable with respect to perturbations in the problem data. A thorough analysis of these processes is given in the book by Lawson and Hanson (1974), where many other aspects of the least squares problem are also examined. A trivial illustration of the advantages of decomposition methods is given in the following example.

Example

$$A = \begin{bmatrix} 1 & 1 \\ 0 & \varepsilon \\ 0 & 0 \end{bmatrix}, \qquad \mathbf{b} = \begin{bmatrix} 1 \\ 1 \\ 1 \end{bmatrix}.$$

If ε is such that $1 + \varepsilon^2$ is represented as 1 on the computer being used, then clearly A^TA is represented as a singular matrix, and the direct approach of solving the normal equations breaks down. In the Golub method, we have $W = I$; the minimum value of the sum of squares is immediately given as 1 and the solution is obtained by solving

$$\begin{bmatrix} 1 & 1 \\ 0 & \varepsilon \end{bmatrix} \mathbf{a} = \begin{bmatrix} 1 \\ 1 \end{bmatrix}.$$

Exercises

1. If A is an $n \times n$ matrix with rank n, show that A^TA is positive definite. If A satisfies the Haar condition, and D is a positive semi-definite diagonal matrix with at least n nonzero elements, prove that A^TDA is positive definite.

2. Prove the parallelogram law

$$\|\mathbf{s}+\mathbf{t}\|^2 + \|\mathbf{s}-\mathbf{t}\|^2 = 2\|\mathbf{s}\|^2 + 2\|\mathbf{t}\|^2.$$

3. Prove that if C is an $m \times n$ matrix

$$\max_{\|\mathbf{x}\|=1} \|C\mathbf{x}\| = \text{(maximum eigenvalue of } C^T C)^{\frac{1}{2}}$$

$$\min_{\|\mathbf{x}\|=1} \|C\mathbf{x}\| = \text{(minimum eigenvalue of } C^T C)^{\frac{1}{2}}.$$

4. Derive the inequality (4.5).

5. Prove Lemma 4.1.

6. For the second worked example in the text, show that the condition number of A is approximately $2/\varepsilon$.

7. Let A be an $(n+1) \times n$ matrix with rank n, let $\mathbf{a}_2$ solve (4.1) with $p = 2$ and let $\mathbf{s} = \mathbf{r}(\mathbf{a}_2)$. Let $\mathbf{a}^*$ minimize $\|\mathbf{r}(\mathbf{a})\|_\infty$ and let $\mathbf{r}^* = \mathbf{r}(\mathbf{a}^*)$. Prove that
   (i) $\text{sign}(r_i^*) = \text{sign}(s_i)$, $i = 1, 2, \ldots, n+1$, $s_i \neq 0$
   (ii) $\|\mathbf{r}^*\|_\infty = \|\mathbf{s}\|_2^2 / \|\mathbf{s}\|_1$.
Illustrate this result for the problems of exercises 3 and 11, Chapter 2. What is the consequence for the L_∞ solution of $s_i = 0$, for some i?

4.3 The cases $p \neq 2$

When p is not equal to 2, then the system of equations (4.3) is no longer linear in the components of $\mathbf{a}$, and thus a solution cannot generally be obtained in a finite number of operations. The system of nonlinear equations can of course be treated using standard methods available for such problems. If $|r_i|^{p-2}$ exists for all i (which is always the case when $p \geq 2$) then if D is defined to be the $m \times m$ diagonal matrix with (i, i) element $|r_i|^{p-2}$, the system (4.3) is just

$$A^T D\mathbf{r} = \mathbf{0}$$

or

$$A^T D A\mathbf{a} = A^T D\mathbf{b}. \tag{4.6}$$

This generalized system of normal equations differs from equations (4.4) by the presence of the weighting matrix D. Most algorithms which have been proposed for solving (4.6) are essentially variants of Newton's method, and recent procedures of this type and appropriate analyses have been given by Fletcher, Grant, and Hebden (1971), Kahng (1972), Ekblom (1973), Merle and Spath (1974), and Wolfe (1979). When $p < 2$, some of the diagonal elements of D, and thus the system (4.6), may not be defined. However, in practice the substitution of small nonzero values for values of r_i may be made, and justified by the finite precision of computer arithmetic. Other devices, for example solving a slightly perturbed problem (Ekblom, 1973) or fixing zero values and so removing the corresponding equations from the original set (Wolfe, 1979), are possible: to simplify the discussion which follows, we will merely assume that D can always be obtained.

Consider the system of equations (4.6) as

$$\mathbf{f}(\mathbf{a}) = \mathbf{0}.$$

Then if Newton's method is applied to this system, the increment $\mathbf{d}_i$ to be added to the current approximation (provided at the ith iteration) $\mathbf{a}_i$ is given by solving

$$\nabla \mathbf{f}(\mathbf{a}_i)\mathbf{d}_i = -\mathbf{f}(\mathbf{a}_i)$$

or

$$(p-1)A^T D_i A\mathbf{d}_i = \mathbf{f}(\mathbf{a}_i),$$

where D_i is a diagonal matrix with (j, j) element $|r_j(\mathbf{a}_i)|^{p-2}$. Thus $\mathbf{d}_i$ satisfies

$$(p-1)\, A^T D_i A\mathbf{d}_i = A^T D_i \mathbf{r}_i, \tag{4.7}$$

writing $\mathbf{r}(\mathbf{a}_i)$ as $\mathbf{r}_i$. In fact $\mathbf{d}_i$ satisfying (4.7) may be obtained through the solution to the problem of finding $\mathbf{z} \in R^n$ to minimize the L_2 norm of

$$\mathbf{s}_i = D_i^{\frac{1}{2}}\mathbf{b} - D_i^{\frac{1}{2}}A\mathbf{z}. \tag{4.8}$$

A point $\mathbf{z}_i$ minimizing $\|\mathbf{s}_i\|_2$ satisfies

$$\begin{aligned} A^T D_i A\mathbf{z}_i &= A^T D_i \mathbf{b} \\ &= A^T D_i(\mathbf{r}_i + A\mathbf{a}_i) \\ &= A^T D_i \mathbf{r}_i + A^T D_i A\mathbf{a}_i. \end{aligned}$$

Thus

$$A^T D_i A (\mathbf{z}_i - \mathbf{a}_i) = A^T D_i \mathbf{r}_i,$$

and it follows that we can take

$$\mathbf{d}_i = \frac{1}{p-1}\,(\mathbf{z}_i - \mathbf{a}_i).$$

Newton's method for solving (4.6) may therefore conveniently be given as follows:

(i) Set $i = 0$, $D_0 = I$.
(ii) Find $\mathbf{z}_i \in R^n$ to minimize $\|\mathbf{s}_i\|_2$.
(iii) Set $\mathbf{a}_{i+1} = \mathbf{a}_i + (\mathbf{z}_i - \mathbf{a}_i)/(p-1)$, where $\mathbf{a}_0 = \mathbf{z}_0$.
(iv) Unless the iteration has converged, set $i = i + 1$, and go to step (ii).

If the matrix $A^T D_i A$ is nonsingular, then there will be a unique $\mathbf{d}_i$ satisfying (4.7). A sufficient condition for this to hold is now given.

Lemma 4.2 Let $\|\mathbf{r}\| > 0$. Then if A satisfies the Haar condition, there exists a neighbourhood of the solution $\mathbf{a}^*$ to (4.1) in which $A^T DA$ is positive definite.

Proof It is a consequence of (4.3) that at least $(n+1)$ components of $\mathbf{r}$ are nonzero at the solution if A satisfies the Haar condition and $\|\mathbf{r}\| \neq 0$. By continuity, this must hold in a neighbourhood of the solution. Thus A^TDA has rank n in this neighbourhood, and the result follows. $\qquad\Box$

Although the Haar condition assumption is not a necessary one for positive definiteness of A^TDA, it is not possible to guarantee this result if the Haar condition is relaxed.

Example

$$A = \begin{bmatrix} 1 & 1 & 1 \\ 1 & 1 & 1 \\ 1 & 0 & 0 \\ 0 & 1 & 0 \end{bmatrix}, \qquad \mathbf{b} = \begin{bmatrix} 1 \\ 2 \\ 0 \\ 0 \end{bmatrix}.$$

It is readily verified that the solution (for all p) is $a_1 = a_2 = 0$, $a_3 = 1.5$, giving $\mathbf{r} = (-0.5, 0.5, 0, 0)^T$.

If the matrices A^TD_iA remain nonsingular, then convergence will be obtained to the solution $\mathbf{a}^*$ of (4.1) provided the starting point (in the algorithm given this is the L_2 solution of (4.1)) is sufficiently close to $\mathbf{a}^*$. It is possible to estimate whether or not the region of convergence of Newton's method is reasonably large, and this turns out to depend on, among other things, the particular value of p. In fact, the size of the domain of convergence may be shown to be roughly proportional to $(p-1)/(p-2)$ (see exercise 8). Thus the Newton method as described is likely to converge for values of p close to 2, but not, for example, for values of p close to 1. It is possible to modify the Newton iteration by making use of the fact that the Newton direction $\mathbf{d}_i$ is a descent direction for $\|\mathbf{r}\|^p$. For, by Taylor expansion

$$\|\mathbf{r}(\mathbf{a}_i + \gamma\mathbf{d}_i)\|^p = \|\mathbf{r}(\mathbf{a}_i)\|^p - \gamma p\mathbf{r}_i^T D_i A \mathbf{d}_i + O(\gamma^2) \qquad (4.9)$$

$$= \|\mathbf{r}(\mathbf{a}_i)\|^p - \frac{\gamma p}{p-1}\mathbf{r}_i^T D_i A (A^T D_i A)^{-1} A^T D_i \mathbf{r}_i + O(\gamma^2)$$

$$< \|\mathbf{r}(\mathbf{a}_i)\|^p$$

for $\gamma > 0$ sufficiently small, if $\mathbf{a}_i \neq \mathbf{a}^*$. Thus we can modify Newton's method so that we take $\mathbf{a}_{i+1} = \mathbf{a}_i + \gamma\mathbf{d}_i$, where γ is chosen so that $\|\mathbf{r}(\mathbf{a}_i)\|^p$ is sufficiently reduced. It has been observed that the choice $\gamma = p - 1$ can be effective for values of $p < 2$, and this is consistent with the observation above on the relative size of the domains of convergence, for when p is close to 1, extremely small steps in the Newton direction will be taken. It is interesting that this particular choice of γ is equivalent to solving (4.6) by a simple iterative scheme in which approximations $\mathbf{a}_{i+1}$ are related to $\mathbf{a}_i$ through the linear system

$$A^T D_i A \mathbf{a}_{i+1} = A^T D_i \mathbf{b}. \qquad (4.10)$$

The verification of this is left to exercise 9.

An appropriate value of γ can also be obtained by methods which guarantee convergence. For example, γ may be chosen at each step as the largest member of the set $\{1, \beta, \beta^2, \ldots\}$ for some $0 < \beta < 1$ (say $\beta = \frac{1}{2}$) such that

$$\psi(\mathbf{a}_i, \gamma) = \frac{\|\mathbf{r}(\mathbf{a}_i)\|^p - \|\mathbf{r}(\mathbf{a}_i + \gamma \mathbf{d}_i)\|^p}{\gamma p \mathbf{r}_i^T D_i A \mathbf{d}_i} \geqslant \sigma > 0 \qquad (4.11)$$

where $\sigma < \frac{1}{2}$ is a constant independent of i, usually chosen to be small, for example 0.0001. That this choice of γ is always possible follows from (4.9), since, for fixed i,

$$\psi(\mathbf{a}_i, \gamma) = 1 + O(\gamma).$$

The quotient $\psi(\mathbf{a}_i, \gamma)$ corresponds to the ratio of the actual decrease in $\|\mathbf{r}\|^p$ to that predicted by the first two terms of the Taylor expansion, and the satisfaction of (4.11) ensures that there is a sufficiently large decrease in the value of the norm to ensure convergence to the solution. The proof of the following theorem draws on analysis given by Fletcher, Grant, and Hebden (1971) and Goldstein and Price (1967).

Theorem 4.3 Let A have rank n, and let $A^T D A$ be positive definite in a neighbourhood B of $\mathbf{a}^*$, the solution to (4.1), defined by

$$\|\mathbf{r}(\mathbf{a})\|^p \leqslant \|\mathbf{r}(\mathbf{a}_0)\|^p.$$

Then

(i) $\mathbf{a}_i \to \mathbf{a}^*$ as $i \to \infty$,
(ii) eventually the choice $\gamma = 1$ satisfies (4.11).

Proof (i) Assume first that the sequence of values of γ, say $\{\gamma_i\}$ is bounded away from zero by $\hat{\gamma}$. Then (4.11) gives

$$\mathbf{r}_i^T D_i A \mathbf{d}_i \leqslant \frac{1}{p\sigma\hat{\gamma}} (\|\mathbf{r}(\mathbf{a}_i)\|^p - \|\mathbf{r}(\mathbf{a}_i + \gamma \mathbf{d}_i)\|^p).$$

The right hand side of this inequality represents the difference between consecutive terms of a monotonically decreasing sequence bounded below, and so tends to zero. Thus

$$\mathbf{r}_i^T D_i A \mathbf{d}_i \to 0 \qquad \text{as} \qquad i \to \infty$$

which, using (4.7), shows that

$$A^T D_i \mathbf{r}_i \to \mathbf{0} \quad \text{as} \quad i \to \infty \qquad (4.12)$$

by the positive definiteness assumption.

Now assume that there exists a subsequence of values γ_i (which we do not rename) tending to zero. Then there exists a subsequence of values $\bar{\gamma}_i = \gamma_i/\beta$ tending to zero and such that

$$\psi(\mathbf{a}_i, \bar{\gamma}_i) < \sigma$$

by the way γ_i is chosen. Now let

$$G(\mathbf{a}) = (p-1)A^T D(\mathbf{a})A.$$

Then Taylor expansion gives

$$\|\mathbf{r}(\mathbf{a}_i + \gamma \mathbf{d}_i)\|^p = \|\mathbf{r}(\mathbf{a}_i)\|^p - \gamma p \mathbf{r}_i{}^T D_i A \mathbf{d}_i + \frac{\gamma^2}{2} p \mathbf{d}_i{}^T G(\xi_i) \mathbf{d}_i, \qquad (4.13)$$

where ξ_i lies between $\mathbf{a}_i$ and $\mathbf{a}_i + \gamma \mathbf{d}_i$, and so

$$1 - \frac{\bar{\gamma}_i \mathbf{d}_i{}^T G(\xi_i) \mathbf{d}_i}{\mathbf{r}_i{}^T D_i A \mathbf{d}_i} < \sigma.$$

Thus

$$\mathbf{r}_i{}^T D_i A \mathbf{d}_i < \frac{\bar{\gamma}_i}{1-\sigma} \mathbf{d}_i{}^T G(\xi_i) \mathbf{d}_i \leqslant \frac{\bar{\gamma}_i W \|\mathbf{d}_i\|_2{}^2}{1-\sigma}$$

by the boundedness of G, where W is constant. Since $\|\mathbf{d}_i\|_2$ is bounded, we must have the result (4.12) as before.

The proof of (i) is now concluded by observing that a consequence of (4.12) is that $\|\mathbf{d}_i\|_2 \to 0$ as $i \to \infty$ so that $\mathbf{a}_i \to \mathbf{a}^*$ as $i \to \infty$.

(ii) From (4.13), we have

$$\psi(\mathbf{a}_i, \gamma) = 1 - \frac{\gamma}{2} \frac{\mathbf{d}_i{}^T G(\xi_i) \mathbf{d}_i}{\mathbf{d}_i{}^T G(\mathbf{a}_i) \mathbf{d}_i}$$

$$= 1 - \frac{\gamma}{2} - \frac{\gamma}{2} \frac{\mathbf{d}_i{}^T [G(\xi_i) - G(\mathbf{a}_i)] \mathbf{d}_i}{\mathbf{d}_i{}^T G(\mathbf{a}_i) \mathbf{d}_i}.$$

Now

$$\mathbf{d}_i{}^T G(\mathbf{a}_i) \mathbf{d}_i \geqslant K \|\mathbf{d}_i\|_2^2,$$

where $K > 0$ is a constant independent of i, by assumption. Thus

$$\left| \psi(\mathbf{a}_i, \gamma) - 1 + \frac{\gamma}{2} \right| \leqslant \frac{\gamma \|G(\xi_i) - G(\mathbf{a}_i)\|}{2K}$$

where the norm on the right hand side is an appropriate matrix norm. Since $\|\xi_i - \mathbf{a}_i\|_2 \to 0$ as $i \to \infty$ the required result follows. $\qquad \square$

Corollary The convergence rate of the method is ultimately second order.

Exercises

8. Show that the size of the domain of convergence of Newton's method applied to (4.6) is essentially proportional to $(p-1)/(p-2)$.

9. If $\mathbf{a}_{i+1}$ is defined by (4.10), show that

$$\mathbf{a}_{i+1} = \mathbf{a}_i + (p-1)\mathbf{d}_i$$

where $\mathbf{d}_i$ satisfies (4.7).

10. Write a program to implement Newton's method for the problems of this section. By considering a simple problem for different values of p, illustrate the result of exercise 8.

11. Extend the program of exercise 10 to incorporate the step length procedure given by (4.11), and apply the method to the problem defined by

$$A = \begin{bmatrix} 1.1 & 0.9 \\ 1.2 & 1.0 \\ 1.0 & 1.0 \end{bmatrix}, \qquad \mathbf{b} = \begin{bmatrix} 2.2 \\ 2.3 \\ 2.1 \end{bmatrix}.$$

Let p become closer and closer to 1. What happens to the values of the components of $\mathbf{r}$ at the solution? Can you give the consequences of this for the solution to the corresponding L_1 problem?

12. Use the Polya algorithm to obtain the best Chebyshev approximation to x^2 by $a_1 x + a_2 e^x$ on 101 equispaced points in $[0, 2]$.

13. The function $\Psi(\mathbf{a}_i, \gamma)$ can be redefined to be the ratio of the actual decrease in $\|\mathbf{r}\|^p$ to that predicted by the first *three* terms of the Taylor expansion. Obtain the appropriate inequality to be satisfied by γ in this case.

14. Let A be an $(n+1) \times n$ matrix with rank n, and let $\mathbf{a}_p$ solve (4.1) with $\mathbf{s} = \mathbf{r}(\mathbf{a}_p)$. Let $\mathbf{a}^*$ minimize $\|\mathbf{r}(\mathbf{a})\|_\infty$, and let $\mathbf{r}^* = \mathbf{r}(\mathbf{a}^*)$. Prove that
(i) $\operatorname{sign}(r_i^*) = \operatorname{sign}(s_i)$, $i = 1, 2, \ldots, n+1$, $s_i \neq 0$,
(ii) $\|\mathbf{r}^*\|_\infty = \mathbf{s}^T D_s \mathbf{s} / \mathbf{s}^T D_s \boldsymbol{\theta}$ where $\theta_i = \operatorname{sign}(s_i)$, $i = 1, 2, \ldots, n+1$, and the subscript s indicates evaluation at $\mathbf{s}$.
Illustrate this result for the problems of exercises 3 and 11, Chapter 2.

15. Let A be an $(n+1) \times n$ matrix with rank n, let $\mathbf{a}_2$ solve (4.1) with $p = 2$, and let $\mathbf{s} = \mathbf{r}(\mathbf{a}_2)$. Let $\mathbf{s}^{(p)}$ be the vector in R^m with ith component

$$(|s_i|/\|\mathbf{s}\|_q)^{q-1} \operatorname{sign}(s_i)$$

where $1/p + 1/q = 1$, and let

$$\mathbf{b}^{(p)} = \mathbf{b} - \frac{\mathbf{b}^T \mathbf{s}}{\|\mathbf{s}\|_q} \mathbf{s}^{(p)}.$$

Prove that the minimizer of $\|\mathbf{r}\|_p$ is given explicitly by

$$\mathbf{a}_p = (A^T A)^{-1} A^T \mathbf{b}^{(p)}.$$

(Levitan and Lynn, 1976).

5

Linear L_p approximation of continuous functions

5.1 Introduction

We now consider the continuous analogue of the approximation problems treated in the previous chapter, so that the appropriate space is $C[X]$, where X is an N-dimensional continuum. Let $f(\mathbf{x})$, $\phi_i(\mathbf{x})$, $i = 1, 2, \ldots, n \in C[X]$, and, as usual, define

$$r(\mathbf{x}, \mathbf{a}) = f(\mathbf{x}) - \sum_{j=1}^{n} a_j \phi_j(\mathbf{x}) \qquad \mathbf{x} \in X$$

where it is assumed that the functions $\{\phi_i(\mathbf{x})\}$ are linearly independent on X. Then we consider the problem

$$\text{find } \mathbf{a} \in R^n \text{ to minimize } \|r(\mathbf{x}, \mathbf{a})\| = \left(\int_X |r(\mathbf{x}, \mathbf{a})|^p dx \right)^{1/p} \tag{5.1}$$

where p satisfies $1 < p < \infty$. The most natural setting for problems of this type is in fact not the space $C[X]$ but the space of functions whose pth power is integrable on X. However, since most commonly occurring functions are continuous, there is no real loss in restricting consideration to $C[X]$ and thereby maintaining consistency with the other approximation problems treated in this book. The relationship between the class of problems (5.1) and those defined by (4.1) is not as clear as in the corresponding Chebyshev approximation problems. However, not surprisingly, the theory, and also the methods, introduced in Chapter 4 are not substantially different from most of what appears here. As usual existence of best approximations is guaranteed, as is uniqueness of the solution to (5.1) by the strict convexity of the normed linear space and the linear independence assumption. We turn therefore to the characterization of the solution, and have the following result.

Theorem 5.1 A vector $\mathbf{a} \in R^n$ solves (5.1) if and only if

$$\int_X \text{sign}(r(\mathbf{x}, \mathbf{a})) |r(\mathbf{x}, \mathbf{a})|^{p-1} \phi_j(\mathbf{x}) dx = 0 \qquad j = 1, 2, \ldots, n. \qquad (5.2)$$

(If $\|r\| = 0$, the result is trivial, so assume not.)

Proof 1 In the notation of Theorem 1.7, $v(\mathbf{x}) \in V(r(\mathbf{x}, \mathbf{a}))$ if and only if it is the unique element defined by

$$v(\mathbf{x}) = \text{sign}(r(\mathbf{x}, \mathbf{a})) |r(\mathbf{x}, \mathbf{a})|^{p-1} \|r(\mathbf{x}, \mathbf{a})\|^{1-p}.$$

The conditions of Theorem 1.7 immediately give (5.2).

Proof 2 The point minimizing $\|r(\mathbf{x}, \mathbf{a})\|$ also minimizes $\|r(\mathbf{x}, \mathbf{a})\|^p$, which is a differentiable function of $\mathbf{r}$, and therefore of $\mathbf{a}$. Thus the minimum occurs at a point satisfying (5.2). If $\mathbf{a}$ satisfies (5.2), then since $\|r(\mathbf{x}, \mathbf{a})\|^p$ is a convex function of $\mathbf{a}$, exercise 15 of Chapter 1 shows that $\mathbf{a}$ solves (5.1), and this concludes the proof. $\qquad \square$

For the problem (5.1), the assumption that the set $\{\phi_i(\mathbf{x})\}$ forms a Chebyshev set on X does not have the same significance, either theoretically or practically, that it has in the case of Chebyshev approximation. However, it is a unifying assumption in the sense of the following interpolation property.

Theorem 5.2 Let $X = [a, b]$ and $\phi_i(x)$, $i = 1, 2, \ldots, n$ be a Chebyshev set on $[a, b]$. Let $\mathbf{a}$ solve (5.1) with $\|r(x, \mathbf{a})\| > 0$. Then $r(x, \mathbf{a})$ changes sign at least n times in $[a, b]$.

Proof Let $r(x, \mathbf{a})$ change sign at $s < n$ points in $[a, b]$, and let the points $a = x_0 < x_1 < \ldots < x_n = b$ contain among them all such points. Now (5.2) gives

$$\sum_{i=1}^{n} \theta_i \int_{x_{i-1}}^{x_i} |r(x, \mathbf{a})|^{p-1} \phi_j(x) dx = 0 \qquad j = 1, 2, \ldots, n,$$

where $\theta_i = \pm 1$, or

$$G^T \theta = 0$$

where G is the $n \times n$ matrix with (i, j) element $\int_{x_{1-i}}^{x_i} |r(x, \mathbf{a})|^{p-1} \phi_j(x) dx$. Thus G is singular, and so there exists $\mathbf{d} \in R^n$, $\mathbf{d} \neq 0$, such that $G\mathbf{d} = 0$. In other words

$$\sum_{j=1}^{n} d_j \int_{x_{i-1}}^{x_i} |r(x, \mathbf{a})|^{p-1} \phi_j(x) dx = 0 \qquad i = 1, 2, \ldots, n$$

or

$$\int_{x_{i-1}}^{x_i} \sum_{j=1}^{n} d_j |r(x, \mathbf{a})|^{p-1} \phi_j(x) dx = 0 \qquad i = 1, 2, \ldots, n.$$

It follows that the integrand in these equations has a zero in each interval (x_{i-1}, x_i), $i = 1, 2, \ldots, n$. This violates the Chebyshev set assumption, and proves the result. □

Exercise

1. Let $X = [a, b]$, and let $\phi_i(x) = x^{i-1}$. Give a direct proof of Theorem 5.2 in this case.

5.2 The case $p = 2$

When $p = 2$, (5.2) can be written

$$\int_X r(\mathbf{x}, \mathbf{a})\phi_j(\mathbf{x})dx = 0 \qquad j = 1, 2, \ldots, n$$

or

$$\mathbf{Ma} = \mathbf{c} \tag{5.3}$$

where M is the $n \times n$ matrix with (i, j) element $\int_X \phi_i(\mathbf{x})\phi_j(\mathbf{x})dx$ and $\mathbf{c} \in R^n$ has $c_i = \int_X \phi_i(\mathbf{x})f(\mathbf{x})dx$, $i = 1, 2, \ldots, n$. Equations (5.3) correspond precisely to the normal equations in the discrete least squares case, and have a unique solution provided that M is nonsingular. In fact, it may be shown (exercise 2) that M is positive definite. However, as in the discrete case, the matrix of the normal equations can be ill-conditioned with respect to inversion.

Example $X = [0, 1]$, $\phi_i = x^{i-1}$, $i = 1, 2, \ldots, n$

$$\int_X \phi_i\phi_j dx = \int_0^1 x^{i-1}x^{j-1}dx = \frac{1}{i+j-1} \qquad i, j = 1, 2, \ldots, n.$$

$$M = \begin{bmatrix} 1 & \dfrac{1}{2} & \dfrac{1}{3} & \cdots & \dfrac{1}{n} \\ \dfrac{1}{2} & \dfrac{1}{3} & \dfrac{1}{4} & \cdots & \dfrac{1}{n+1} \\ \vdots & & & & \\ \dfrac{1}{n} & \dfrac{1}{n+1} & & \cdots & \dfrac{1}{2n-1} \end{bmatrix}.$$

This is the upper left hand corner of the well-known Hilbert matrix, and is notoriously ill-conditioned with respect to inversion. For $n = 10$, M^{-1} has elements as large as 3×10^{12}. Thus errors of 10^{-10} in $\mathbf{c}$ (for example, rounding errors) could give rise to errors as large as 300 in the components of $\mathbf{a}$.

There is no method for this problem which corresponds to the orthogonal factorization method recommended for the discrete least squares problem. The only way in which difficulties of this kind may be avoided is by different choice of the basis functions $\phi_i(\mathbf{x})$, $i = 1, 2, \ldots, n$. For example, if these were such that $\int_X \phi_i(\mathbf{x}) \phi_j(\mathbf{x}) \, dx = 0$ when $i \neq j$, then M would be a diagonal matrix, and problems directly associated with the inversion of M could be entirely avoided. In this case, the basis functions are said to be *orthogonal*. For convenience, we introduce the inner product notation defined by

$$\langle \phi_i, \phi_j \rangle = \int_X \phi_i(\mathbf{x}) \phi_j(\mathbf{x}) \, dx. \tag{5.4}$$

Then the solution to the least squares problem when the basis functions are orthogonal can be written down explicitly as

$$a_i = \langle \phi_i, f \rangle / \langle \phi_i, \phi_i \rangle \qquad i = 1, 2, \ldots, n.$$

If, in addition to orthogonality, the basis functions are normalized so that $\langle \phi_i, \phi_i \rangle = 1$, i.e. are *orthonormal*, then of course $M = I$ and we have

$$a_i = \langle \phi_i, f \rangle \qquad i = 1, 2, \ldots, n.$$

It is common to refer to these coefficients as the *Fourier coefficients* of $f(x)$.

Exercises

2.   Prove that the matrix M defined in the previous section is positive definite if the functions $\{\phi_i(\mathbf{x})\}$ are linearly independent.

3.   Find the best L_2 approximation to x^2 in $[0, 1]$ by a linear combination of (i) 1 and x, (ii) x^{100} and x^{101}. Compare the answers.

4.   Let $\phi_i(x)$ denote a polynomial of degree $(i - 1)$ in x. Determine $\phi_1(x)$, $\phi_2(x)$, $\phi_3(x)$ so that these polynomials are mutually orthogonal in $[-1, 1]$. Hence find the best L_2 approximation to $\cos \pi x$ on $[-1, 1]$ by a polynomial of degree 2.

5.   Let $f(x) \in C[0, 1]$ and let $\phi(x)$ satisfy

$$\phi(x) = \begin{cases} x/\xi & 0 \leqslant x \leqslant \xi \\ (1-x)/(1-\xi) & \xi \leqslant x \leqslant 1. \end{cases}$$

Let $e(\xi, a) = \int_0^1 (f(x) - a\phi(x))^2 \, dx$ where ξ is assumed variable. Show that $e(\xi, a)$ has two stationary values given by
   (i) $a = 0$
   (ii) $a = 3 \displaystyle\int_0^1 f(x) \phi(x) \, dx$, with ξ being chosen so that with this value of a, $a\phi(x)$
   provides the best L_2 approximation to $f(x)$ *simultaneously* on $[0, \xi]$ and $[\xi, 1]$.
Show that the minimum of $e(\xi, a)$ is attained when (ii) holds.

5.3 Orthogonal functions

Any linearly independent system of functions may be used to generate an orthogonal (or in fact orthonormal) system of functions by use of the *Gram–Schmidt process*. Let $p_1, p_2, \ldots, p_n$ be a linearly independent set. Then choose

$$q_1 = p_1/\|p_1\|,$$
$$q_j = u_j/\|u_j\| \tag{5.5}$$

where

$$u_j = p_j - \sum_{i=1}^{j-1} \langle p_j, q_i \rangle q_i \qquad j = 2, 3, \ldots, n$$

and the norm is the L_2 norm.

Theorem 5.3 The system of functions $q_j, j = 1, 2, \ldots, n$, defined by (5.5) is orthonormal.

Proof We use induction. Clearly $\{q_1\}$ is orthonormal, since p_1 cannot be zero. Assume that $\{q_1, q_2, \ldots, q_k\}$ are orthonormal. Then

$$\langle u_{k+1}, q_j \rangle = \langle p_{k+1}, q_j \rangle - \sum_{i=1}^{k} \langle p_{k+1}, q_i \rangle \langle q_i, q_j \rangle \qquad j = 1, 2, \ldots, k$$
$$= \langle p_{k+1}, q_j \rangle - \langle p_{k+1}, q_j \rangle \qquad j = 1, 2, \ldots, k$$
$$= 0 \qquad j = 1, 2, \ldots, k.$$

Thus

$$\langle q_{k+1}, q_j \rangle = 0 \qquad j = 1, 2, \ldots, k$$

and, further, $\langle q_{k+1}, q_{k+1} \rangle = 1$. Thus by the induction assumption, $\{q_1, q_2, \ldots, q_n\}$ is an orthonormal system. $\qquad\square$

Orthogonal systems of functions may also be defined with respect to a positive weight function $w(x)$, that is a function $w(x)$ satisfying $w(x) > 0$, $x \in X$. We generalize the inner product notation to redefine

$$\langle \phi_i, \phi_j \rangle = \int_X w(x)\, \phi_i(x) \phi_j(x)\, dx. \tag{5.6}$$

Then the argument of Theorem 5.3 goes through as before for this inner product, where the norm is replaced by the *weighted least squares norm*

$$\|r(x)\|_w = \left(\int_X w(x) |r(x)|^2\, dx \right)^{\frac{1}{2}} = \langle r, r \rangle^{\frac{1}{2}}. \tag{5.7}$$

It follows that for the problem of finding $a \in R^n$ to minimize $\|r(x, a)\|_w$, where the basis functions $\phi_i(\mathbf{x})$ satisfy

$$\langle \phi_i, \phi_j \rangle = \delta_{ij} = \begin{cases} 0 & i \neq j \\ 1 & i = j \end{cases}$$

we immediately have the solution given by

$$a_i = \langle \phi_i, f \rangle \qquad i = 1, 2, \ldots, n$$

the *generalized Fourier coefficients* of $f(x)$. Notice that the solution is such that its components are independent of n; this is in complete contrast to the L_∞ case.

The study of orthogonal functions and their properties is not something which lies in the mainstream of this book. Nevertheless, it is of sufficient importance, both theoretically and practically, for us to devote some space to it. In particular, we will consider the important special case of orthogonal polynomials. Since whole books have been published on this topic, and indeed on particular systems of such functions (for example, Szego, 1939; Fox and Parker, 1968; Rivlin, 1974), we can only give a brief introduction to some of the more important aspects.

5.4 Orthogonal Polynomials

In this section, we will assume that $X = [a, b]$. Then, given a weight function $w(x)$ satisfying $w(x) > 0$, $x \in [a, b]$, a system of orthogonal polynomials can be generated by the Gram–Schmidt process applied to the set of independent functions $1, x, x^2, \ldots$. Such polynomials have certain properties which make them particularly convenient to work with, and some of these are now described. Firstly, it is clear that if $\phi_0(x)$, $\phi_1(x)$, $\ldots$, $\phi_n(x)$ is such a system of orthogonal polynomials then $\phi_i(x)$ is a polynomial of degree i. (Notice that we have altered the subscripts so that this fact is reflected; in this case the coefficient vector will be in R^{n+1}, and not in R^n as before.) We now give two important properties of such systems.

Theorem 5.4 The orthogonal polynomials $\phi_0(x)$, $\ldots$, $\phi_n(x)$ are linearly independent, and any arbitrary polynomial of degree $r \leqslant n$ can be uniquely expressed as a linear combination of $\phi_0(x)$, $\ldots$, $\phi_r(x)$.

Proof For any $r \leqslant n$,

$$\int_a^b w(x)\phi_r(x)\left\{ \sum_{i=0}^n a_i\phi_i(x) \right\} dx = a_r \langle \phi_r, \phi_r \rangle.$$

Thus $\sum_{i=0}^n a_i\phi_i(x) = 0$ implies that $a_r = 0$, $r = 0, 1, \ldots, n$.

For the second part, clearly $\phi_0(x)$ is a constant and so the result holds for $r = 0$. Assume that the result is true for r, and let $\sum_{i=0}^{r+1} d_i x^i$ be an arbitrary polynomial of degree $\leqslant r+1$. Then

$$\sum_{i=0}^{r+1} d_i x^i = d_{r+1} x^{r+1} + \sum_{i=0}^{r} d_i x^i.$$

Further

$$\phi_{r+1}(x) = b_{r+1} x^{r+1} + \sum_{i=0}^{r} b_i x^i \qquad \text{where } b_{r+1} \neq 0$$

and so

$$\sum_{i=0}^{r+1} d_i x^i = \frac{d_{r+1}}{b_{r+1}} \phi_{r+1}(x) + \sum_{i=0}^{r} \lambda_i \phi_i(x)$$

where the coefficients on the right hand side are unique. The result for general $r \leqslant n$ follows by induction. $\qquad\qquad\square$

Theorem 5.5 The orthogonal polynomials $\phi_0(x), \ldots, \phi_n(x)$ satisfy the three-term recurrence relation

$$\phi_{r+1}(x) = (\alpha_r x + \beta_r) \phi_r(x) + \gamma_{r-1} \phi_{r-1}(x) \qquad r = 1, 2, \ldots, n-1 \qquad (5.8)$$

where α_r, β_r, and γ_{r-1} are independent of x.

Proof We can choose α_r so that $\phi_{r+1}(x) - \alpha_r x \phi_r(x)$ is a polynomial of degree r, and is therefore (by Theorem 5.4) uniquely expressible in the form

$$\phi_{r+1}(x) - \alpha_r x \phi_r(x) = \sum_{i=0}^{r-1} \gamma_i \phi_i(x) + \beta_r \phi_r(x),$$

say. Multiplying by $w(x)\phi_j(x)$ and integrating gives

$$\langle \phi_j, \phi_{r+1} - \alpha_r x \phi_r \rangle = \gamma_j \langle \phi_j, \phi_j \rangle \qquad j = 0, 1, \ldots, r-1, \qquad (5.9)$$

and similarly

$$\langle \phi_r, \phi_{r+1} - \alpha_r x \phi_r \rangle = \beta_r \langle \phi_r, \phi_r \rangle.$$

But

$$\begin{aligned}
\langle \phi_j, \phi_{r+1} - \alpha_r x \phi_r \rangle &= \langle \phi_j, \phi_{r+1} \rangle - \alpha_r \langle \phi_j, x \phi_r \rangle && j = 0, 1, \ldots, r \\
&= 0 - \alpha_r \langle x \phi_j, \phi_r \rangle && j = 0, 1, \ldots, r \\
&= -\alpha_r \left\langle \sum_{i=0}^{j+1} d_i \phi_i, \phi_r \right\rangle && j = 0, 1, \ldots, r.
\end{aligned}$$

The right hand side of this equation is zero for $j = 0, 1, \ldots, r-2$ and so, using (5.9),

$$\gamma_j = 0 \qquad j = 0, 1, \ldots, r-2.$$

We therefore have the required result. $\qquad\qquad\square$

Once the coefficients α_r, β_r, and γ_{r-1} have been determined for a particular case, we thus have a convenient way of obtaining systems of orthogonal polynomials. Note that this result is valid for polynomials only, and a similar result does not hold in general for other orthogonal functions. The question arises, then, of how we can determine the particular values of α_r, β_r, and γ_{r-1}. Now, unless the polynomials are normalized, we are free to choose any scaling, so let us make the assumption that the leading coefficients are unity, in other words that the orthogonal polynomials are *monic*. Then we must have $\alpha_r = 1$, and the recurrence relation (5.8) becomes

$$\phi_{r+1}(x) = (x + \beta_r)\phi_r(x) + \gamma_{r-1}\phi_{r-1}(x) \qquad r = 1, 2, \ldots, n-1. \quad (5.10)$$

Taking inner products with $\phi_r(x)$ and $\phi_{r-1}(x)$ respectively gives

$$\beta_r = -\frac{\langle x\phi_r, \phi_r \rangle}{\langle \phi_r, \phi_r \rangle} \qquad (5.11)$$

and

$$\gamma_r = -\frac{\langle x\phi_r, \phi_{r-1} \rangle}{\langle \phi_{r-1}, \phi_{r-1} \rangle}. \qquad (5.12)$$

The use of (5.10), (5.11), (5.12) (starting with $\phi_0 = 1$, $\phi_1(x) = x + \beta_0$) enables any system of orthogonal polynomials to be generated. For the interval $[-1, 1]$, an important general class is that of polynomials orthogonal with respect to the weight function $(1 + x)^{\lambda_1}(1 - x)^{\lambda_2}$, defined for $\lambda_1, \lambda_2 > -1$. These are known as *Jacobi polynomials*, and particular choices of the parameters λ_1 and λ_2 give rise to many important examples. Some of these, and also some others, are given in Table 2.

Table 2

X	$w(x)$	Polynomial name
$[-1, 1]$	$(1 + x)^{\lambda_1}(1 - x)^{\lambda_2}\lambda_1, \lambda_2 > -1$	Jacobi
$[-1, 1]$	$(1 - x^2)^{\lambda - \frac{1}{2}}$	Gegenbauer
$[-1, 1]$	$(1 - x^2)^{-\frac{1}{2}}$	Chebyshev (1st kind) $T_n(x)$
$[-1, 1]$	$(1 - x^2)^{\frac{1}{2}}$	Chebyshev (2nd kind) $U_n(x)$
$[-1, 1]$	1	Legendre $P_n(x)$
$(-\infty, \infty)$	e^{-x^2}	Hermite $H_n(x)$
$[0, \infty)$	$x^\alpha e^{-x}, \alpha > -1$	Generalized Laguerre $L_n^\alpha(x)$

Of special interest is the set of Legendre polynomials, since these are the appropriate basis functions for the original L_2 problem (with weight function 1). The recurrence relation can in this case be given in the particularly simple form

$$P_{r+1}(x) = xP_r(x) - \frac{r^2}{(2r + 1)(2r - 1)}P_{r-1}(x); \qquad (5.13)$$

the first 5 (monic) Legendre polynomials are

$$P_0 = 1$$
$$P_1 = x$$
$$P_2 = x^2 - \tfrac{1}{3}$$
$$P_3 = x^3 - \tfrac{3}{5}x$$
$$P_4 = x^4 - \tfrac{6}{7}x^2 + \tfrac{3}{35}.$$

Legendre polynomials may also be defined explicitly, through Rodrigues' formula (see exercise 21). Indeed it is frequently the case that orthogonal polynomials can be defined in a convenient alternative way. This is the case for the important class of orthogonal polynomials known as the Chebyshev polynomials of the first kind, which may be defined by means of the formula

$$T_r(x) = \cos r\theta \qquad r = 0, 1, \ldots, n, \tag{5.14}$$

where $x = \cos\theta$. The orthogonality property can be verified directly by making a change of variable from x to θ and making use of properties of integrals of products of trigonometric functions. We have

$$\int_{-1}^{1} T_k(x) T_m(x) \frac{dx}{\sqrt{(1-x^2)}} = 0 \qquad m \neq k \tag{5.15a}$$

$$\int_{-1}^{1} T_k(x)^2 \frac{dx}{\sqrt{(1-x^2)}} = \begin{cases} \pi/2 & k \neq 0 \\ \pi & k = 0. \end{cases} \tag{5.15b}$$

It may also be readily verified that the appropriate three-term recurrence relation for $T_r(x)$ is

$$T_{r+1}(x) = 2x T_r(x) - T_{r-1}(x) \qquad r = 1, 2, \ldots, n-1, \tag{5.16}$$

with $T_0(x) = 1$, $T_1(x) = x$. The scaling introduced by the definition (5.14) is such that for $r \geq 2$, $T_r(x)$ is not monic; in fact the coefficient of x^r is easily seen to be 2^{r-1}, and so the actual form of (5.16) differs from that of (5.10).

The Chebyshev polynomials $T_r(x)$ have certain advantages over other orthogonal polynomials for numerical work, and they are often used as basis functions in approximation problems. In order to pursue this, we first introduce the concept of *orthogonal expansions*. For *any* system of orthogonal polynomials, the *expansion of $f(x)$* in terms of these functions may be written

$$\sum_{i=0}^{\infty} a_i \phi_i(x), \tag{5.17}$$

where

$$a_i = \frac{\langle \phi_i, f \rangle}{\langle \phi_i, \phi_i \rangle} \qquad i = 0, 1, \ldots .$$

This expansion has the property that its partial sums $p_n(x)$ are best degree n polynomial approximations to $f(x)$ in the appropriate weighted least squares sense. Let $p_n*(x)$ denote the best L_∞ approximation to $f(x)$ by a polynomial of degree n in $[a, b]$. Then

$$\|f(x) - p_n(x)\|_w^2 \leqslant \|f(x) - p_n*(x)\|_w^2$$

$$\leqslant \int_a^b w(x)(f(x) - p_n*(x))^2 \, dx$$

$$\leqslant \max_{a \leqslant x \leqslant b} |f(x) - p_n*(x)|^2 \int_a^b w(x) \, dx$$

$$= E_n^2(f) \int_a^b w(x) \, dx$$

using the notation introduced in Section 3.7. It follows from Weierstrass' theorem that

$$\|f(x) - p_n(x)\|_w \to 0 \quad \text{as} \quad n \to \infty,$$

in other words we obtain *convergence in the mean*. However, in general $p_n(x)$ does not converge uniformly to $f(x)$ as $n \to \infty$, so that (5.17) may not necessarily be regarded as a *representation* of $f(x)$. For some orthogonal polynomials, uniform convergence may however be proved under reasonably mild conditions, as the following result illustrates.

Theorem 5.6 Let $f(x) \in C^2[-1, 1]$. Then the expansion of $f(x)$ in terms of Chebyshev polynomials $T_i(x)$ converges uniformly to $f(x)$.

Proof The Chebyshev expansion is

$$\sum_{i=0}^{\infty} a_i T_i(x)$$

where

$$a_0 = \frac{1}{\pi} \int_{-1}^1 \frac{f(x)}{\sqrt{(1-x^2)}} \, dx$$

$$a_i = \frac{2}{\pi} \int_{-1}^1 \frac{f(x) T_i(x)}{\sqrt{(1-x^2)}} \, dx \qquad i = 1, 2, \ldots.$$

For any $j \geqslant 1$, we have

$$a_j = \frac{2}{\pi} \int_0^\pi f(\cos\theta) \cos j\theta \, d\theta$$

$$= \frac{2}{\pi} \int_0^\pi F(\theta) \cos j\theta \, d\theta \quad \text{where } F(\theta) = f(\cos\theta),$$

$$= -\frac{2}{\pi j} \int_0^\pi DF(\theta) \sin j\theta \, d\theta, \quad \text{integrating by parts,}$$

$$= \frac{2}{\pi j^2} DF(\theta)[\cos j\pi - 1] - \frac{2}{\pi j^2} \int_0^\pi D^2 F(\theta) \cos j\theta \, d\theta,$$

integrating by parts once again.

Thus for constant M,

$$|a_j| \leqslant Mj^{-2} \qquad j = 1, 2, \ldots \qquad (5.18)$$

and so there exists $g(x) \in C[-1, 1]$ such that

$$\|p_n(x) - g(x)\|_\infty \to 0 \quad \text{as } n \to \infty$$

where $p_n(x)$ is the nth partial sum of the Chebyshev expansion. Now

$$\|f(x) - g(x)\|_w \leqslant \|f(x) - p_n(x)\|_w + \|p_n(x) - g(x)\|_w$$

showing that $f(x) \equiv g(x)$, and concluding the proof. $\qquad \square$

For *any* (uniformly) convergent expansion (5.17), where the polynomials are normalized so that $|\phi_i(x)| \leqslant 1$ in $[a, b]$,

$$f(x) - p_n(x) = \sum_{i=n+1}^{\infty} a_i \phi_i(x)$$

$$\leqslant \sum_{i=n+1}^{\infty} |a_i|.$$

so that if the coefficients a_i for $i > n$ are negligible, $p_n(x)$ is a satisfactory representation of $f(x)$. The most rapidly convergent expansion will be the one whose coefficients a_i go to zero fastest, and the inequality (5.18) suggests that the Chebyshev expansion is a likely candidate. If it is assumed that a_n is inversely proportional to the coefficient of x^n in $\phi_n(x)$, then of all convergent series, the most rapidly convergent will be the one with the *largest* coefficient of x^n in $\phi_n(x)$, subject to the normalization that $|\phi_n(x)| \leqslant 1$ in $[-1, 1]$. This turns out to be indeed the Chebyshev expansion, as a consequence of the following so-called *minimax property*. (See also Rivlin and Wilson, 1969.)

Theorem 5.7 Of all monic polynomials of degree n, the Chebyshev polynomial $2^{1-n}T_n(x)$ is the best Chebyshev approximation to zero in $[-1, 1]$.

Proof We have

$$2^{1-n}T_n(x) = 2^{1-n}\cos n\theta$$

and so the maximum value in $[-1, 1]$ is 2^{1-n}. Differentiating with respect to θ and setting to zero gives as turning points those points such that

$$-2^{1-n}n \sin n\theta = 0$$

or

$$\theta = \frac{r\pi}{n} \qquad r = 0, 1, \ldots, n.$$

Thus $2^{1-n}T_n(x)$ attains its maximum (modulus) value at $(n+1)$ points in $[-1, 1]$

with alternating sign. But this is just the error when $-x^n$ is approximated by a polynomial of degree $(n-1)$, and so the result follows by Theorem 3.3. $\qquad\square$

It is an immediate consequence of Theorem 5.7 that the Chebyshev polynomial $T_n(x)$ has the largest coefficient of x^n of all polynomials of degree n which have a maximum value of unity in $[-1, 1]$. The truncated Chebyshev series $p_n(x)$ is in fact very close to the best degree n polynomial approximation to $f(x)$ in the Chebyshev sense. We will make this more precise, and require the following preliminary lemma.

Lemma 5.1 Let $p_n(x, g)$ be the truncated Chebyshev series expansion to degree n of $g(x)$. Then

$$p_n(x, g) = \frac{1}{\pi} \int_{-\pi}^{\pi} G(\phi) \frac{\sin\left[(n+\frac{1}{2})(\phi+\theta)\right]}{2 \sin(\phi+\theta)/2} \, d\phi, \tag{5.19}$$

where $g(x) = g(\cos\theta) = G(\theta)$, and we define $G(-\theta) = G(\theta)$.

Proof Using the definitions of the Chebyshev series coefficients, we have

$$p_n(x, g) = \frac{2}{\pi} \sum_{j=0}^{n}{}' \int_{-1}^{1} \frac{g(y) T_j(y) T_j(x)}{\sqrt{(1-y^2)}} \, dy$$

where the dash indicates that the first term of the sum is halved. Putting $y = \cos\phi$, $x = \cos\theta$, we have

$$p_n(x, g) = \frac{1}{\pi} \sum_{j=0}^{n}{}' \int_{-\pi}^{\pi} G(\phi) \cos j\phi \cos j\theta \, d\phi.$$

Using the evenness of G, and some trigonometric identities, we obtain the required result. The details are left as an exercise. $\qquad\square$

The following result is given by Powell (1967); see also Rivlin (1974).

Theorem 5.8

$$\|f - p_n\|_\infty \leqslant (1 + L_n) E_n(f)$$

where

$$L_n = \frac{1}{\pi} \int_0^{\pi} \left| \frac{\sin(n+\frac{1}{2})\phi}{\sin\frac{1}{2}\phi} \right| \, d\phi.$$

Proof For any $x \in [-1, 1]$,

$$|f(x) - p_n(x)| = |f(x) - p_n{}^*(x) + p_n{}^*(x) - p_n(x)|$$

where $p_n{}^*$ is the best L_∞ polynomial of degree n approximation to $f(x)$. Thus

$$|f(x) - p_n(x)| \leqslant E_n(f) + p_n(x, p_n^* - f).$$

Since we may regard $G(\phi)$ as being periodic, period 2π, the integrand in (5.19) is periodic, period 2π, and so we can write

$$p_n(x, g) = \frac{1}{\pi} \int_{-\pi}^{\pi} G(\phi - \theta) \frac{\sin(n + \frac{1}{2})\phi}{2\sin\frac{1}{2}\phi} \, d\phi$$

$$= \frac{1}{2\pi} \int_0^{\pi} [G(\phi + \theta) + G(\phi - \theta)] \frac{\sin(n + \frac{1}{2})\phi}{\sin\frac{1}{2}\phi} \, d\phi.$$

Thus

$$p_n(x, p_n^* - f) \leqslant E_n(f) L_n$$

and we have the required result. $\qquad\qquad\qquad\qquad\qquad\qquad\qquad\square$

The number L_n defined in the previous theorem is known as the nth *Lebesque constant*. It may be shown (see, for example, Rivlin, 1969) that the inequality

$$L_n < 3 + \frac{4}{\pi^2} \log n \tag{5.20}$$

is satisfied; thus, for values of n up to 1000, $1 + L_n$ does not exceed 5.2, and the loss of accuracy in using p_n instead of p_n^* is small. Notice that this result suggests that for the computation of p_n^* by the second algorithm of Remes, a good initial set of $(n + 1)$ points is the extrema of $T_{n+1}(x)$.

In addition to the use of the truncated Chebyshev expansion as an approximate representation of a function $f(x)$, the Chebyshev polynomials are often used as basis functions when the coefficients are to be found by some other means, for example interpolation. One reason for this is that the coefficients of this representation can again be much smaller than those of the powers of x in the corresponding approximating polynomial, and thus better precision may be maintained.

Finally, we look briefly at the problem of the efficient evaluation of a Chebyshev series at a particular point x (see also exercise 20). Let the series be

$\sum\limits_{r=0}^{n}{}' a_r T_r(x)$ and consider the recurrence relation

$$b_r(x) = 2x b_{r+1}(x) - b_{r+2}(x) + a_r \qquad r = n, n-1, \ldots, 0 \tag{5.21}$$

where $b_{n+1}(x) = b_{n+2}(x) = 0$. Then

$$\sum_{r=0}^{n}{}' a_r T_r = \tfrac{1}{2} a_0 + \sum_{r=1}^{n} a_r T_r$$

$$= \tfrac{1}{2}(b_0 - 2x b_1 + b_2) + \sum_{r=1}^{n} (b_r - 2x b_{r+1} + b_{r+2}) T_r$$

$$= \tfrac{1}{2}(b_0 - 2x b_1 + b_2) + b_1 T_1 + b_2 T_2 - 2x b_2 T_1$$

$$+ \sum_{r=3}^{n} (b_r T_r - 2x b_r T_{r-1} + b_r T_{r-2})$$

$$- 2x b_{n+1} T_n + b_{n+1} T_{n-1} + b_{n+2} T_n.$$

Making use of the recurrence relation (5.16), we obtain

$$\sum_{r=0}^{n} {}' a_r T_r(x) = \tfrac{1}{2}(b_0 - b_2).$$ (5.22)

An error analysis of this process is given by Fox and Parker (1968), who show that the evaluation of the Chebyshev series in this way is a numerically stable calculation. If ε_r is the *local* error in computing b_r from (5.21) (that is, the error in b_r assuming that all other quantities are exact), then the error in (5.22) is bounded above by $\sum_{r=0}^{n} {}' |\varepsilon_r|$.

Exercises

6. Let $\phi_0(x), \ldots, \phi_n(x)$ be any system of orthogonal polynomials with respect to the weight function $w(x)$ in $[a, b]$ and let

$$p_n(x) = \sum_{i=0}^{n} a_i \phi_i(x)$$

be the (weighted) least squares approximation of $f(x) \in C[a, b]$. Prove that:

(i) $\langle f - p_n, \phi_j \rangle = 0, j = 0, 1, \ldots, n$

(ii) $\sum_{i=0}^{n} a_i^2 \langle \phi_i, \phi_i \rangle \leq \langle f, f \rangle$
    (Bessel's inequality)

(iii) $\sum_{i=0}^{\infty} a_i^2 \langle \phi_i, \phi_i \rangle = \langle f, f \rangle$
    (Parseval's formula).

7. The set $\{1, x, x^2 - \tfrac{3}{5}\}$ is orthogonal with respect to $w(x) = x^2$ in $[-1, 1]$. Find the weighted least squares approximation to x^3 on $[-1, 1]$ by a polynomial of degree 2.

8. Prove the following properties of Chebyshev polynomials $T_r(x)$:
    (i) $T_{r+1}(x) + T_{r-1}(x) = 2x T_r(x), r = 1, 2, \ldots$.
    (ii) If r is even (odd), $T_r(x)$ is an even (odd) function.
    (iii) $(1 - x^2) D^2 T_r(x) - x D T_r(x) + r^2 T_r(x) = 0$.
    (iv) $T_{2r}(x) = T_r(2x^2 - 1)$.
    (v) $T_r(T_s(x)) = T_{rs}(x)$

(vi) $\displaystyle\int T_r(x) dx = \begin{cases} T_1 & r = 0 \\ \tfrac{1}{4} T_2 & r = 1 \\ \dfrac{1}{2}\left[\dfrac{T_{r+1}}{r+1} - \dfrac{T_{r-1}}{r-1} \right] & r > 1. \end{cases}$

9. Complete the details of the proof of Lemma 5.1.

10. Put $T_r(x) = \mathcal{R}e^{ir\theta}$ and show that if $|c| < 1$, c real,

$$1 - cT_1(x) + c^2 T_2(x) - c^3 T_3(x) + \ldots = \frac{1}{2} + \frac{1}{2} \frac{1 - c^2}{1 + c^2 + 2cx}.$$

By suitable choice of c, deduce from this a Chebyshev series for $(4x + 5)^{-1}$.

11. For $f(x) = e^x$, compare the errors associated with the truncated Chebyshev expansion up to degree 4 with those for the degree 4 Maclaurin polynomial in the range $-1 \leqslant x \leqslant 1$.

12. Let $q_n(x)$ be the polynomial of degree n which interpolates $f(x) \in C^{n+1}$ $[-1, 1]$ at $(n+1)$ distinct points in $[-1, 1]$. Prove that the error bound introduced in Section 1.6 is minimized if the points x_i are chosen as the zeros of $T_{n+1}(x)$.

13. Consider the differential equation

$$\mathcal{L}(y(x)) = 0 \qquad -1 \leqslant x \leqslant 1,$$

where $\mathcal{L}$ is a differential operator. Let $\phi(x, \mathbf{a})$ be an approximate solution satisfying any boundary conditions, where the components of $\mathbf{a} \in R^n$ are obtained by satisfying

$$\mathcal{L}(\phi(x_i, \mathbf{a})) = 0 \qquad i = 1, 2, \ldots, n$$

for points $x_1, x_2, \ldots, x_n$ in $[-1, 1]$. Justify the choice

$$x_i = \cos \frac{(2i - 1)\pi}{n} \qquad i = 1, 2, \ldots, n.$$

14. Show that

(i) $\displaystyle |x| = \frac{2}{\pi} + \frac{4}{\pi} \sum_{r=1}^{\infty} \frac{(-1)^{r-1}}{4r^2 - 1} T_{2r}(x)$

(ii) $\displaystyle (1 - x^2)^{1/2} = \frac{2}{\pi} - \frac{4}{\pi} \sum_{r=1}^{\infty} \frac{1}{4r^2 - 1} T_{2r}(x).$

Let $J_r(x)$ be the Bessel function of order r, which satisfies

$$J_r(x) = \begin{cases} \dfrac{1}{\pi} \displaystyle\int_0^\pi \cos r\phi \, \cos(x \sin \phi) d\phi & r \text{ even} \\[3mm] \dfrac{1}{\pi} \displaystyle\int_0^\pi \sin r\phi \, \sin(x \sin \phi) d\phi & r \text{ odd}. \end{cases}$$

Show that

(iii) $\displaystyle \cos x = J_0(1) + 2 \sum_{r=1}^{\infty} (-1)^r J_{2r}(1) T_{2r}(x)$

(iv) $\displaystyle \sin x = 2 \sum_{r=0}^{\infty} (-1)^r J_{2r+1}(1) T_{2r+1}(x)$

(v) $e^x = J_0(i) + 2 \sum_{r=1}^{\infty} i^r J_r(-i) T_r(x)$

where $i = \sqrt{(-1)}$.

15. To obtain the truncated Chebyshev expansion, integrals

$$\int_{-1}^{1} (1-x^2) T_r(x) f(x) dx$$

have to be evaluated, and this may be difficult analytically. Obtain the appropriate expansion when these integrals are evaluated numerically using the trapezoidal rule at intervals π/N in θ.

16. The recurrence relation (5.21) may be modified to take account of special features of the expression to be evaluated. Let

$$P(x) = \sum_{i=0}^{n}{}' a_i T_i(x).$$

(i) Assume that $a_{2i+1} = 0$, $i = 0, 1, 2, \ldots, m-1$, where $n = 2m$. Show that we can use the recurrence relation

$$b_r(x) = 2(2x^2 - 1) b_{r+1}(x) - b_{r+2}(x) + a_{2r} \qquad r = m, m-1, \ldots, 0$$
$$b_{m+1}(x) = b_{m+2}(x) = 0$$

and have $P(x) = \frac{1}{2}(b_0 - b_2)$ as before.

(ii) Assume that $a_{2i} = 0$, $i = 0, 1, \ldots, m-1$, where $n = 2m-1$. Show that

$$P(x) = x \sum_{i=0}^{m-1}{}' \bar{a}_{2i} T_{2i}(x)$$

where

$$\bar{a}_{2i} = 2a_{2i+1} - \bar{a}_{2i+2} \qquad i = m-1, m-2, \ldots, 0$$
$$\bar{a}_{2m} = 0,$$

so that $P(x)/x$ may be evaluated as in (i).

17. Assuming the truncated Chebyshev expansion for $\cos x$:

$$0.765198 T_0 - 0.229807 T_2 + 0.004953 T_4 - 0.000042 T_6,$$

derive by integration the corresponding series for $\sin x$, and use it to approximate $\sin (\pi/4)$, using the result of the previous exercise.

18. By considering the binomial expansion of $\cos^n \theta = \{\frac{1}{2}(e^{i\theta} + e^{-i\theta})\}^n$, prove that

$$x^n = 2^{1-n} \sum_{r=0}^{[\frac{1}{2}n]}{}' {}^n C_r T_{n-2r}(x),$$

where $[\frac{1}{2}n]$ denotes the greatest integer less than or equal to $\frac{1}{2}n$, ${}^n C_r$ is a binomial coefficient, and *if n is even* the coefficient of $T_0(x)$ is halved. Deduce that $x^2 = \frac{1}{2}(T_0 + T_2)$, $x^3 = \frac{1}{4}(3T_1 + T_3)$, $x^4 = \frac{1}{8}(3T_0 + 4T_2 + T_4)$, and obtain similar expressions up to x^7.

19. Use the Maclaurin expansion for sin x and the results of the previous example to derive a Chebyshev series to degree 7 for sin x. Calculate an approximation to sin $(\pi/4)$ from this series, and compare with the result obtained in exercise 16.

20. Let $\phi_0(x), \ldots, \phi_n(x)$ be any system of monic orthogonal polynomials, and for given $a_0, a_1, \ldots, a_n$ let

$$P(x) = \sum_{i=0}^{n} a_i \phi_i(x).$$

Consider the recurrence relation

$$b_r(x) = (x + \beta_r)b_{r+1}(x) + \gamma_r b_{r+2}(x) + a_r \qquad r = n, n-1, \ldots, 0,$$

where $b_{n+1}(x) = b_{n+2}(x) = 0$, and β_r and γ_r are given by (5.11) and (5.12). Using the recurrence relation (5.10), prove that $P(x) = b_0$.

21. If the Legendre polynomials are defined by Rodrigues' formula

$$P_n(x) = \frac{1}{2^n n!} D^n\{(x^2 - 1)^n\},$$

establish the relations

(i) $\displaystyle \int_{-1}^{1} P_n(x)P_m(x) = \begin{cases} 2/2n+1 & n = m \\ 0 & n \neq m \end{cases}$

(ii) $(n+1)P_{n+1}(x) = (2n+1)xP_n(x) - nP_{n-1}(x)$.
Obtain the analogue of Theorem 5.7 for the L_2 norm.

22. Show that the system of functions $\{1, \cos x, \sin x, \cos 2x, \ldots\}$ is orthogonal with respect to $w(x) = 1$ in $-\pi \leqslant x \leqslant \pi$. The infinite expansion

$$\tfrac{1}{2}a_0 + \sum_{r=1}^{\infty} (a_r \cos rx + b_r \sin rx)$$

defined by (5.17) gives the *Fourier series* of $f(x)$. If $f(x)$ is 2π-periodic, continuous everywhere, and differentiable at x, then its Fourier series converges at x to $f(x)$ (see, for example, Fox and Parker, 1968).

23. The Chebyshev polynomials of the second kind may be defined by

$$U_n(x) = \frac{\sin (n+1)\theta}{\sin \theta}$$

where $x = \cos \theta$. Show that:
(i) $U_n(x)$ is a polynomial of degree n in x with leading coefficient 2^n.

(ii) $U_{n+1}(x) + U_{n-1}(x) = 2x U_n(x)$.

(iii) $U_n(x) - U_{n-2}(x) = 2T_n(x)$.

(iv) $U_n(x) - x U_{n-1}(x) = T_n(x)$.

(v) $x U_{n-1}(x) - U_{n-2}(x) = T_n(x)$.

24. Let $\phi_0(x), \phi_1(x), \ldots$ be a system of polynomials orthogonal with respect to $w(x)$ on $[a, b]$. Given $f(x) \in C[a, b]$, let $q_n(x, f)$ denote the polynomial of degree n which interpolates to $f(x)$ at the zeros of $\phi_{n+1}(x)$, say $x_0, x_1, \ldots, x_n$. Prove that:

(i) $l_i(x) = \dfrac{\phi_{n+1}(x)}{(x - x_i) D \phi_{n+1}(x_i)}$ $\quad i = 0, 1, \ldots, n$

where $l_i(x)$ is the ith Lagrange coefficient.

(ii) $\langle l_i, l_j \rangle = 0$ $\quad i \ne j$.

(iii) $\displaystyle\sum_{i=0}^{n} \int_a^b l_i^2(x) w(x) dx = \int_a^b w(x) dx$.

(iv) $\| q_n(x, f) - p_n^*(x) \|_w \to 0$ as $n \to \infty$, where p_n^* is the best Chebyshev approximation to $f(x)$ in $[a, b]$ of degree n.

(v) $\| q_n(x, f) - f(x) \|_w \to 0$ as $n \to \infty$.

(The last result is the *Erdös–Turan Theorem*.)

25. Define a *discrete* inner product by

$$\langle f, g \rangle = \sum_{j=0}^{n} {}'' f(x_j) g(x_j)$$

where the double prime means that the first and last terms are halved, and

$$x_j = \cos\frac{j\pi}{n} \quad j = 0, 1, \ldots, n.$$

Prove that the Chebyshev polynomials $T_j(x)$ are orthogonal with respect to this inner product.

26. Let $q_n(x)$ be a polynomial of degree n which satisfies $|q_n(x)| \le 1$, $-1 \le x \le 1$. Show that

$$|q_n(x)| \le |T_n(x)| \quad |x| > 1$$

(Rogosinski, 1955).

27. If $q_n(x)$ is a polynomial of degree n with n distinct zeros in $[-1, 1]$ and $|q_n(\cos k\pi/n)| = 1$, $k = 0, 1, \ldots, n$, show that either $q_n(x) = T_n(x)$ or $q_n(x) = -T_n(x)$ (Devore, 1974). This result is in fact true provided only that the zeros of $q_n(x)$ lie in the set $S = \{x : |1 - T_n^2(x)| \le 1\}$ (Micchelli and Rivlin, 1974).

28. Let $\phi_0(x), \ldots, \phi_n(x)$ be orthogonal with respect to $w(x)$ in $[a, b]$. Prove that

(i) $|\langle f, \phi_n \rangle| \le E_{n-1}(f) \int_a^b |\phi_n(x)| w(x) dx$

(ii) $|\langle f, \phi_n \rangle| \le E_{n-1}(f) (\int_a^b w(x) dx)^2$.

29. Let $p_n(x)$ be the truncated Chebyshev expansion of $f(x)$. Prove that if $f \in C^2[a, b]$,

$$\| f(x) - p_n(x) \|_w^2 \leqslant \frac{\pi^2 \| D^2 f \|_\infty^2}{6(n-1)^3}.$$

(Hint: Use exercise 6, 27, and the inequality (3.24)).

30. Use Theorem 5.8 and the inequalities (5.20) and (3.23) to prove the *Dini–Lipschitz Theorem*: if $f \in C[-1, 1]$ satisfies $\lim_{n \to \infty} (\log n)\, \omega(1/n) = 0$, then $p_n(x)$ converges uniformly to $f(x)$ in $[-1, 1]$ as $n \to \infty$. ($\omega \equiv \omega_0$ is the modulus of continuity of f).

5.5 The cases $p > 2$

Having digressed into the area of orthogonal polynomials and some of their many interesting properties, we now return to the original problem (5.1) for values of p satisfying $p > 2$. Equations (5.2) represent a system of n nonlinear equations in the n unknowns a_i, $i = 1, 2, \ldots, n$, and techniques similar to those described in Chapter 4 are available for obtaining the solution. We can write

$$\int_X r(\mathbf{x}, \mathbf{a})|r(\mathbf{x}, \mathbf{a})|^{p-2}\phi_j(\mathbf{x})\mathrm{d}x = 0 \qquad j = 1, 2, \ldots, n$$

so that

$$M(\mathbf{a})\mathbf{a} = \mathbf{c}(\mathbf{a}) \tag{5.23}$$

where M is the $n \times n$ matrix with (i, j) element

$$\int_X |r(\mathbf{x}, \mathbf{a})|^{p-2}\phi_i(\mathbf{x})\phi_j(\mathbf{x})\mathrm{d}x$$

and $\mathbf{c} \in R^n$ has ith component

$$c_i = \int_X |r(\mathbf{x}, \mathbf{a})|^{p-2}f(\mathbf{x})\phi_i(\mathbf{x})\mathrm{d}x \qquad i = 1, 2, \ldots, n.$$

Thus, comparing with (5.3), we see that the function $|r(\mathbf{x}, \mathbf{a})|^{p-2}$ may be regarded as a weight function which plays a role analogous to that of the matrix D in the discrete case. Notice that if $p < 2$, this function is likely to become infinite at points in X, and the analysis of this problem is not straightforward; we have merely ignored this case. If Newton's method is applied to the system (5.23) at the current approximation $\mathbf{a}_i$, then the increment $\mathbf{d}_i$ satisfies the system of equations

$$(p-1)M(\mathbf{a}_i)\mathbf{d}_i = \mathbf{g}_i, \tag{5.24}$$

where the jth component of $\mathbf{g}_i$ is

$$\int_X |r(\mathbf{x}, \mathbf{a}_i)|^{p-2}r(\mathbf{x}, \mathbf{a}_i)\phi_j(\mathbf{x})\mathrm{d}x.$$

This may be compared directly with equation (4.7). The linear independence of the set $\{\phi_i(x)\}$ on X is sufficient to ensure that $M(\mathbf{a}_i)$ is positive definite if $\|r(\mathbf{x}, \mathbf{a}_i)\| \neq 0$ and thus a unique $\mathbf{d}_i$ satisfying (5.24) always exists. The methods and analysis introduced in Chapter 4 for the corresponding discrete problem carry over directly to this case. The reader is left to verify the details; the paper by Fletcher, Grant, and Hebden (1971) is a useful guide.

Exercises

31.   Prove that the matrix M of equation (5.23) is positive definite.

32.   Let $\mathbf{z}_i$ satisfy

$$M(\mathbf{a}_i)\mathbf{z}_i = \mathbf{c}(\mathbf{a}_i).$$

Prove that

$$\mathbf{d}_i = \frac{1}{p-1}(\mathbf{z}_i - \mathbf{a}_i)$$

satisfies (5.24). Is there any advantage in calculating $\mathbf{d}_i$ in this way?

33.   Let $f(x) \in C^{n+1}[-1, 1]$ and let Π_n be the subspace of polynomials of degree n. Prove that there exists $\xi \in (-1, 1)$ such that

$$\inf_{p \in \Pi_n} \|f - p\|_p = |D^{n+1}f(\xi)|\delta_n^{(p)}/(2^n(n+1)!)$$

where

$$\delta_n^{(p)} = \inf \|(x - x_0)(x - x_1) \ldots (x - x_n)\|_p$$

and the infimum is taken over all x_i such that

$$-1 \leqslant x_0 \leqslant x_1 \leqslant \ldots \leqslant x_n \leqslant 1.$$

(Hint: Use Theorem 5.2 and the expression for the error in the interpolating polynomial; Phillips, 1970.)

34.   Let the right hand side of (5.24) be denoted by $\mathbf{g}(\mathbf{a}_i, p)$, and let $C(\mathbf{a}, p)$ be the $n \times n$ matrix with (j, k) element

$$-\frac{1}{(p-1)}\frac{\partial}{\partial a_k}g_j(\mathbf{a}, p).$$

Show that if $\mathbf{a}_p$, the solution to (5.1), is considered as a function of p, it is continuous and differentiable, with the derivative of the jth component given by

$$\frac{1}{(p-1)}\sum_{k=1}^{n}[C(\mathbf{a}_p, p)]_{jk}^{-1}\int_X |r(\mathbf{x}, \mathbf{a}_p)|^{p-2}r(\mathbf{x}, \mathbf{a}_p)\log|r(\mathbf{x}, \mathbf{a}_p)|\phi_k(\mathbf{x})d\mathbf{x}.$$

(Fletcher, Grant, and Hebden, 1974b).

6

The L_1 solution of an overdetermined system of linear equations

6.1 Introduction

This chapter is concerned with approximation problems set in R^m normed with the L_1 norm, a criterion which appears to originate with Laplace (1799). In particular, we consider the problem

$$\text{find } \mathbf{a} \in R^m \text{ to minimize } \|\mathbf{r}(\mathbf{a})\| = \sum_{i=1}^{m} |r_i(\mathbf{a})|, \qquad (6.1)$$

where $\mathbf{r}(\mathbf{a}) = \mathbf{b} - A\mathbf{a}$, and as usual A is an $m \times n$ matrix and $\mathbf{b} \in R^m$. The L_1 norm is important in the analysis of data which includes gross inaccuracies or blunders (wild points), for the effective weight given to such values is smaller than with any other L_p norm (see Barrodale, 1968). A simple illustration of this is given in Figure 10 where we show the results of fitting a straight line to the 4 points A(0, 0), B(1, 1), C(2, 3), and D(3, 3) in the (x, y) plane using the L_1, L_2, and L_∞ norms. For the L_1 solution, we need to minimize

$$\|\mathbf{r}\| = |a_1| + |a_1 + a_2 - 1| + |a_1 + 2a_2 - 3| + |a_1 + 3a_2 - 3|,$$

where the straight line is $y = a_1 + a_2 x$. This expression defines a convex region in R^3 whose vertices correspond to points $\mathbf{a}$ where $r_j = r_k = 0$, for pairs of indices j and k. The best L_1 approximation is in fact given by $\mathbf{a} = (0, 1)^T$, corresponding to the line $y = x$ in Figure 10. For the L_∞ norm, the best approximation is the line $y = \frac{1}{2} + x$ and for the L_2 norm, $y = 1/10 + 11x/10$. Now, under the assumption that a linear model is appropriate, the point C is likely to be in error, and thus the L_1 solution is the best of the three. Indeed the position of the point C does not affect the L_1 solution, whereas it plays an essential part in defining the best L_∞ approximation. The L_1 solution is the only one of the three which actually

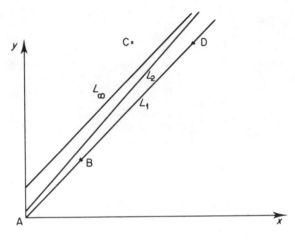

Figure 10

interpolates to any of the data, and we will see that this property of interpolation is an important aspect of the characterization of best L_1 approximations. Claerbout and Muir (1973) describe the situation in the following way: if, on arriving at a fork in the road, the route to be taken is decided by setting up and solving a best approximation problem, then the L_1 solution will result in one of the two roads being chosen, while the L_2 solution will lead off into the bushes.

Exercise

1.   Obtain graphical solutions to the problems of exercise 1 of Chapter 2.

6.2   Characterization of solutions

Characterization results for (6.1) may be given in different forms, of which the following is perhaps the most convenient. We will use $Z = Z(\mathbf{a})$ to denote the set of indices i for which $r_i(\mathbf{a}) = 0$, and $V = V(\mathbf{a})$ to denote the set

$$V(\mathbf{a}) = \{\mathbf{v} : \mathbf{v} \in R^m, \ |v_i| \leqslant 1; \ v_i = \text{sign}\,(r_i(\mathbf{a})), \ i \notin Z\}.$$

Theorem 6.1   A vector $\mathbf{a} \in R^n$ solves (6.1) if and only if there exists $\mathbf{v} \in V$ such that

$$A^T\mathbf{v} = \mathbf{0}. \tag{6.2}$$

Proof 1   In the notation of Theorem 1.7, $V(\mathbf{r})$ is identical to the set V defined here, and so the result is immediate.

Proof 2   Let the conditions of the theorem be satisfied at $\mathbf{a}$. Then

$$\|\mathbf{r}(\mathbf{a})\| = \mathbf{v}^T\mathbf{r} = \mathbf{v}^T\mathbf{b}.$$

For any $\mathbf{d} \in R^n$

$$\|\mathbf{r}(\mathbf{d})\| = \sum_{i=1}^{m} |r_i(\mathbf{d})| \geqslant \sum_{i=1}^{m} v_i r_i(\mathbf{d}) = \mathbf{v}^T \mathbf{b}.$$

Thus $\mathbf{a}$ solves (6.1).

Now let $\mathbf{a}$ be a solution to (6.1) and assume that the conditions of the theorem are not satisfied. Let

$$D = \{\mathbf{d} \in R^n : \mathbf{d} = A^T\mathbf{v}, \mathbf{v} \in V\}.$$

Then D is a closed, convex subset of R^n (exercise 2), with $\mathbf{0} \notin D$. Thus, by Theorem 1.5, there exists $\mathbf{c} \in R^n$ such that for all $\mathbf{d} \in D$, $\mathbf{d}^T\mathbf{c} > 0$, so that

$$\mathbf{v}^T A\mathbf{c} > 0 \qquad \text{for all } \mathbf{v} \in V. \tag{6.3}$$

Now (recalling that $\boldsymbol{\alpha}_i^T$ denotes the ith row of A),

$$\|\mathbf{r}(\mathbf{a} + \gamma\mathbf{c})\| = \sum_{i=1}^{m} |r_i(\mathbf{a} + \gamma\mathbf{c})|$$

$$= \sum_{i=1}^{m} |r_i(\mathbf{a}) - \gamma\boldsymbol{\alpha}_i^T\mathbf{c}|$$

$$= \sum_{i \notin Z} v_i(r_i(\mathbf{a}) - \gamma\boldsymbol{\alpha}_i^T\mathbf{c}) + \gamma \sum_{i \in Z} |\boldsymbol{\alpha}_i^T\mathbf{c}|,$$

for $\gamma > 0$ sufficiently small, say $\gamma < G$, for which sign $r_i(\mathbf{a} + \gamma\mathbf{c})$ remains constant equal to sign $r_i(\mathbf{a})$ for all $i \notin Z$. Choose

$$v_i = -\text{sign}\,(\boldsymbol{\alpha}_i^T\mathbf{c}) \qquad i \in Z.$$

Then if $0 < \gamma < G$,

$$\|\mathbf{r}(\mathbf{a} + \gamma\mathbf{c})\| = \|\mathbf{r}(\mathbf{a})\| - \gamma\mathbf{v}^T A\mathbf{c}$$
$$< \|\mathbf{r}(\mathbf{a})\|$$

using (6.3). This contradicts the assumption that $\mathbf{a}$ solves (6.1) and proves the result. $\qquad\qquad\square$

The aforementioned interpolation property of best L_1 approximations is quantified in the following theorem.

Theorem 6.2 If A has rank t, a solution $\mathbf{a}$ to (6.1) always exists with Z containing at least t indices.

Proof Let A have rank t, and let $\mathbf{a}$ solve (6.1) with $Z(\mathbf{a})$ containing $s < t$ indices. Let $\mathbf{c}$, $\|\mathbf{c}\|_2 = 1$, satisfy

$$\boldsymbol{\alpha}_i^T\mathbf{c} = 0 \qquad i \in Z \tag{6.4}$$

and let $\mathbf{v}$ satisfy (6.2). Then for $|\gamma| > 0$ sufficiently small

$$
\begin{aligned}
\|\mathbf{r}(\mathbf{a} + \gamma \mathbf{c})\| &= \sum_{i=1}^{m} |r_i(\mathbf{a}) - \gamma \boldsymbol{\alpha}_i{}^T \mathbf{c}| \\
&= \sum_{i \notin Z} v_i(r_i(\mathbf{a}) - \gamma \boldsymbol{\alpha}_i{}^T \mathbf{c}) \\
&= \|\mathbf{r}(\mathbf{a})\| - \gamma \sum_{i \notin Z} v_i \boldsymbol{\alpha}_i{}^T \mathbf{c} \\
&= \|\mathbf{r}(\mathbf{a})\|
\end{aligned}
$$

Now there exists $\mathbf{c}$ satisfying (6.4) with $\boldsymbol{\alpha}_j{}^T \mathbf{c} \neq 0$ for some $j \notin Z$, by the rank assumption. Thus $|\gamma|$ may be increased away from zero until the first component of $r_i(\mathbf{a} + \gamma \mathbf{c})$, $i \notin Z$ becomes zero while $\mathbf{a} + \gamma \mathbf{c}$ is still a solution of (6.1). The number of zero components can be increased in this way until no nontrivial solution to (6.4) is possible, when Z contains at least t indices. $\qquad\square$

Corollary The submatrix of A consisting of the rows of A corresponding to $Z(\mathbf{a})$ must have rank t for some solution $\mathbf{a}$ to (6.1).

Theorem 6.3 Let S denote the set of solutions to (6.1), and let K denote the convex hull of all $\mathbf{a} \in S$ for which $Z(\mathbf{a})$ contains t indices. Then $S \equiv K$.

Proof Clearly $K \subset S$. Let $\mathbf{a}^* \in S$, but $\mathbf{a}^* \notin K$. Then the closed, convex set $\{\mathbf{k} - \mathbf{a}^*, \mathbf{k} \in K\}$ does not contain the origin, and so by Theorem 1.5 there exists $\mathbf{u} \in R^n$ such that for all $\mathbf{k} \in K$,

$$
\mathbf{u}^T \mathbf{k} < \mathbf{u}^T \mathbf{a}^* = \beta, \text{ say.}
$$

Now for any $\mathbf{a} \in S$, with $Z(\mathbf{a})$ containing $s < t - 1$ indices, we can choose $\mathbf{c} \in R^n$, $\|\mathbf{c}\|_2 = 1$, such that

$$
\begin{aligned}
\boldsymbol{\alpha}_i{}^T \mathbf{c} &= 0 \qquad i \in Z(\mathbf{a}), \\
\mathbf{u}^T \mathbf{c} &= 0.
\end{aligned}
$$

Thus, as in Theorem 6.2, we can find $\mathbf{d} \in S$ with $Z(\mathbf{d})$ containing $(t-1)$ indices and satisfying $\mathbf{u}^T \mathbf{d} = \beta$. Let $\mathbf{c} \in R^n$, $\|\mathbf{c}\|_2 = 1$, be such that

$$
\begin{aligned}
\boldsymbol{\alpha}_i{}^T \mathbf{c} &= 0 \qquad i \in Z(\mathbf{d}), \\
\mathbf{u}^T \mathbf{c} &\geqslant 0.
\end{aligned}
$$

Then by increasing γ away from zero in a positive direction, we can obtain $\mathbf{d} + \gamma \mathbf{c} \in S$ with $Z(\mathbf{d} + \gamma \mathbf{c})$ containing t indices. Further, $\mathbf{u}^T(\mathbf{d} + \gamma \mathbf{c}) \geqslant \beta$, which gives a contradiction. It follows that $S \subset K$ and the proof is complete. $\qquad\square$

The previous result shows that if A has rank n (the usual case) it is possible to compute a best approximation by finding all vectors $\mathbf{a}$ which are defined by setting n components of $\mathbf{r}$ to zero. It is interesting that the importance of such subsets of n indices reflects the dual nature of the L_1 and L_∞ norms: in the L_∞ case we saw the corresponding importance of subsets of $(n+1)$ indices where the components of $\mathbf{r}$ had the same absolute value. Before dealing in detail with computational methods for the L_1 problem, however, we make brief mention of uniqueness.

There do not appear to exist any simple conditions on the data of the problem (6.1) which will guarantee the uniqueness of the solution. In view of some results which will be shown to hold for the continuous L_1 problem (Chapter 7) it might be expected that an appropriate condition would be the Haar condition on A. However, this is not the case as the following example illustrates.

Example

$$A = \begin{bmatrix} 1 \\ -1 \end{bmatrix}, \quad \mathbf{b} = \begin{bmatrix} 1 \\ 2 \end{bmatrix}.$$

It is readily seen that the minimum value of $\|\mathbf{r}\|$ occurs for all values of a_1 satisfying $-2 \leqslant a_1 \leqslant 1$, yet A satisfies the Haar condition. Notice that the particular solutions $a_1 = -2$ and $a_1 = 1$ are the interpolative ones predicted by Theorem 6.2.

Exercises

2. Show that the set D of Theorem 6.1 is convex.

3. An alternative statement to that of Theorem 6.1 is the following: $\mathbf{a}$ solves (6.1) if and only if

$$\sum_{i \in Z} |\alpha_i{}^T \mathbf{d}| \geqslant |\sum_{i \notin Z} \text{sign}\,(r_i(\mathbf{a}))\alpha_i{}^T \mathbf{d}|$$

for all $\mathbf{d} \in R^n$. Verify this, and also obtain this result directly.

4. Determine all the vertices of $\|\mathbf{r}\|$, and hence solve (6.1) for the cases

(i)
$$A = \begin{bmatrix} 1 & 2 \\ 0 & 1 \\ 1 & 2 \end{bmatrix}, \quad \mathbf{b} = \begin{bmatrix} 1 \\ 2 \\ 0 \end{bmatrix}$$

(ii)
$$A = \begin{bmatrix} 1 & 0 \\ 2 & 0 \\ 3 & 0 \end{bmatrix}, \quad \mathbf{b} = \begin{bmatrix} 2 \\ -1 \\ 1 \end{bmatrix}.$$

5. Let S denote the set of solutions to (6.1). For any i, $1 \leqslant i \leqslant m$, show that the set $\{r_i(\mathbf{a}), \mathbf{a} \in S\}$ cannot contain both positive and negative elements.

6. Let A have rank n, and let $\mathbf{a}^{(p)}$ minimize $\|\mathbf{r}(\mathbf{a})\|_p$. Prove that (c.f. exercise 14 of Chapter 1)

(i) $\|\mathbf{r}(\mathbf{a})^{(p)}\|_p \to \|\mathbf{r}(\mathbf{a}^*)\|_1$ where $\mathbf{a}^*$ solves (6.1),

(ii) if $\mathbf{a}^*$ is unique, $\{\mathbf{a}^{(p)}\} \to \mathbf{a}^*$.

(In view of the corresponding result as $p \to \infty$, it might be expected that $\{\mathbf{a}^{(p)}\}$ always converges as $p \to 1$; however, this is still an open question.)

7.   Prove that if the solution to (6.1) is unique, it is strongly unique. (Hint: let $\gamma = \min_{\|\mathbf{c}\|_2 = 1} \max_{\mathbf{v} \in V} \mathbf{v}^T A \mathbf{c}$, and show first that $\gamma > 0$.)

8.   Let A be an $(n+1) \times n$ matrix with rank n. Let $\mathbf{a}_2$ minimize $\|\mathbf{r}\|_2$, let $\mathbf{s} = \mathbf{r}(\mathbf{a}_2)$, and define the set S of indices

$$S = \{i : |s_i| = \|\mathbf{s}\|_\infty\}$$

Prove that the minimum of $\|\mathbf{r}\|_1$ is unique if and only if S contains just one index (Duris and Sreedharan, 1968).

9.   Let A be an $m \times n$ matrix with rank n, let $\mathbf{r}(\mathbf{b}, \mathbf{a}) = \mathbf{b} - A\mathbf{a}$, and let $Z(\mathbf{b}, \mathbf{a})$ denote the set of indices i such that $r_i(\mathbf{b}, \mathbf{a}) = 0$. For given $\mathbf{b} = \mathbf{b}^{(1)}$, let $\mathbf{a}^{(1)}$ minimize $\|\mathbf{r}(\mathbf{b}^{(1)}, \mathbf{a})\|_1$ with $Z(\mathbf{b}^{(1)}, \mathbf{a}^{(1)})$ containing at least n indices, and let $Z = \{\sigma_1, \sigma_2, \ldots, \sigma_n\}$ be a subset of n of these indices such that the corresponding $n \times n$ submatrix of A is nonsingular. For given $\mathbf{b} = \mathbf{b}^{(2)}$, let $\mathbf{a}^{(2)}$ be uniquely defined by the equations

$$r_i(\mathbf{b}^{(2)}, \mathbf{a}^{(2)}) = 0 \qquad i \in Z,$$

and let

$$\Delta(\mathbf{b}, j) = \det \begin{bmatrix} \boldsymbol{\alpha}_{\sigma_1}^T & b_{\sigma_1} \\ \cdot & \cdot \\ \cdot & \cdot \\ \cdot & \cdot \\ \boldsymbol{\alpha}_{\sigma_n}^T & b_{\sigma_n} \\ \boldsymbol{\alpha}_j^T & b_j \end{bmatrix}.$$

Then if
  (i) $Z(\mathbf{b}^{(2)}, \mathbf{a}^{(2)}) \supset Z(\mathbf{b}^{(1)}, \mathbf{a}^{(1)})$,
  (ii) there exists $\theta = \pm 1$ such that, for any j, $1 \leq j \leq m$, either sign $(\Delta(\mathbf{b}^{(2)}, j)) = \theta$ sign $(\Delta(\mathbf{b}^{(1)}, j))$ or $\Delta(\mathbf{b}^{(2)}, j) = 0$,

prove that $\mathbf{a}^{(2)}$ minimizes $\|\mathbf{r}(\mathbf{b}^{(2)}, \mathbf{a})\|_1$ (Ascher, 1978).

6.3 Linear programming methods

The first systematic methods for solving (6.1) were essentially geometric in nature, originating from work of Edgeworth in the 1880s. The observation that the linear discrete L_1 approximation problem could be formulated as a linear programming problem was made by Wagner (1959), and this opened the way for the solution of large general problems of the type (6.1) in an efficient manner. The special structure inherent in the problem was exploited by Barrodale and Young (1966) and Barrodale (1970), and recent improvements in linear programming based methods have been concerned with modifying the simplex method to accelerate the calculation of the optimal solution. The linear programming formulation of (6.1) requires that $\mathbf{r}$ be replaced by

$$r_i = u_i - v_i \qquad i = 1, 2, \ldots, m$$

where $u_i, v_i \geq 0$, $i = 1, 2, \ldots, m$. Also let

$$a_j = c_j - d_j \qquad j = 1, 2, \ldots, n$$

where $c_j, d_j \geq 0, j = 1, 2, \ldots, n$. Now consider the following linear programming problem

$$\text{minimize} \sum_{i=1}^{m} (u_i + v_i)$$

subject to

$$\mathbf{u} - \mathbf{v} + A(\mathbf{c} - \mathbf{d}) = \mathbf{b}$$
$$\mathbf{u}, \mathbf{v}, \mathbf{c}, \mathbf{d} \geq 0.$$

This may be restated as

$$\text{minimize} \begin{bmatrix} \mathbf{e}^T & \mathbf{e}^T & \mathbf{0}^T & \mathbf{0}^T \end{bmatrix} \begin{bmatrix} \mathbf{u} \\ \mathbf{v} \\ \mathbf{c} \\ \mathbf{d} \end{bmatrix}$$

$$\text{subject to} \begin{bmatrix} I & -I & A & -A \end{bmatrix} \begin{bmatrix} \mathbf{u} \\ \mathbf{v} \\ \mathbf{c} \\ \mathbf{d} \end{bmatrix} = \mathbf{b} \tag{6.6}$$

$$\mathbf{u}, \mathbf{v}, \mathbf{c}, \mathbf{d} \geq 0.$$

This problem is in standard form, and a solution may be obtained by direct application of the simplex method. Now, at a basic feasible solution, and in particular at the optimal basic feasible solution, the nonsingularity of the basis matrix means that no u_k and corresponding v_k (or c_k and corresponding d_k) can both be present as basic variables. Thus it follows that (in particular)

$$u_i v_i = 0 \qquad i = 1, 2, \ldots, m$$

so that

$$u_i + v_i = |u_i - v_i| = |r_i|.$$

Thus, the problem (6.6) is precisely equivalent to that of minimizing $\|\mathbf{r}\|$, and an optimal value of $\mathbf{a}$ can be immediately obtained from the linear programming solution.

As an alternative to (6.6), the *dual* problem may of course be formulated and solved, giving rise to an *ascent* method. This was originally the recommendation of several authors; however, it is now accepted that there is no advantage to be gained in treating the problem in this way, and we do not pursue this further.

The simplex method may be directly applied to (6.6) and a solution obtained. An initial basic feasible solution is immediately available: if $b_i > 0$, then $\mathbf{e}_i$ can be present in the initial basis matrix, and if $b_i < 0$, then $-\mathbf{e}_i$ can be, with either being available if $b_i = 0$. Thus there is no requirement for the addition of artificial variables, and so no associated first phase problem. Most recent work on this method has been concerned with making good use of the special structure involved in (6.6) to complete efficiently the solution to the problem. Spyropoulos, Kiountouzis, and Young (1973) give an algorithm appropriate when A has rank n, and a modified version which applies when the discrete problem arises from polynomial approximation, which it is claimed is particularly fast and uses minimum storage requirements. Another recent method is due to Barrodale and Roberts (1973); we describe this approach.

Rather than allowing the simplex method to proceed to optimality in the usual way, Barrodale and Roberts achieve efficiency by imposing certain restrictions on the choice of variable to become basic, and also by modifying the rule by which the variable to be made nonbasic is chosen. For the first n iterations (at most), only columns from A or $-A$ are permitted to enter the basis matrix, and only those from I or $-I$ are permitted to leave. The usual simplex rule for the variable to enter the basis is otherwise used, but the variable to leave is chosen so that there is the greatest possible reduction in the objective function. Thus feasibility of a variable u_j (say) may be lost, but can easily be recovered by exchanging u_j with v_j. This is clear, because if u_j is basic with value ϕ, then making v_j basic and u_j nonbasic gives v_j the value $-\phi$. In addition the change in the rth column (say) of the basis matrix from $\mathbf{e}_j$ to $-\mathbf{e}_j$ simply alters the signs of the elements in the rth row of the inverse.

When this procedure can no longer be continued, the current basic feasible solution corresponds to an approximation with Z containing k indices, where k pairs u_i and v_i have value zero. The rank of A is thus $t \leqslant k$, where t is the total number of columns from A and $-A$ in the basis. Usually we will have $t = k$. The next stage involves iterating in such a way that at each step we have a basic feasible solution with Z containing t indices. Thus only columns from I and $-I$ are interchanged, the usual simplex rule being (otherwise) applied to determine the vector to enter the basis, but again that leaving the basis chosen so that there is the greatest possible reduction in the objective function. On conclusion of this stage, some of the variables c_j or d_j may be negative: the basic solution becomes feasible (and thus optimal) by interchanging such variables c_j (or d_j) with the corresponding nonbasic variables d_j (or c_j). The process is precisely equivalent to that involving the interchange of u_j and v_j as described earlier. Further details of the implementation of the method and a Fortran program are given by Barrodale and Roberts (1972). We illustrate by a simple example.

Example

$$A = \begin{bmatrix} 1 & -1 \\ 0 & 1 \\ 1 & 1 \end{bmatrix}, \quad \mathbf{b} = \begin{bmatrix} -1 \\ 2 \\ 4 \end{bmatrix}.$$

To avoid the introduction of linear programming notation, for the purposes of the example we simply express the basic variables in terms of the nonbasic ones. It should be remembered that only the coefficients of $\mathbf{c}$ (say) need actually be transformed and stored, as those of $\mathbf{d}$ can be immediately obtained. Denoting the objective function by z, the initial basic feasible solution is given by

$$
\begin{aligned}
v_1 &= 1 + u_1 && + c_1 - c_2 - d_1 + d_2 \\
u_2 &= 2 + v_2 && \qquad - c_2 \quad + d_2 \\
u_3 &= 4 && + v_3 - c_1 - c_2 + d_1 + d_2 \\
z &= 7 + 2u_1 + 2v_2 + 2v_3 && \quad - 3c_2 \quad + 3d_2.
\end{aligned}
$$

Clearly, c_2 is the variable to become basic, and the usual simplex step would be to make v_1 nonbasic. If c_2 is increased beyond the value 1, then v_1 will become negative, but feasibility can be recovered by exchanging v_1 for u_1. If this were done, the coefficient of c_2 in z would become $(-3) - 2 \times (-1) = -1$, which, being negative, means that a further decrease in z is possible. Thus we write

$$
\begin{aligned}
u_1 &= -1 + v_1 && - c_1 + c_2 + d_1 - d_2 \\
z &= 5 + 2v_1 + 2v_2 + 2v_3 - 2c_1 - c_2 + 2d_1 + d_2,
\end{aligned}
$$

the other two equations being unchanged. Now if c_2 is further increased beyond the value 2, when u_2 becomes zero, we would require to interchange u_2 for v_2 in the basis. The coefficient of c_2 in z would then become $(-1) - 2 \times (-1) = 1$, so that in fact no further decrease is possible. Thus we complete the modified simplex step by interchanging the variables c_2 and u_2. We have then

$$
\begin{aligned}
u_1 &= 1 - u_2 + v_1 + v_2 && - c_1 + d_1 \\
c_2 &= 2 - u_2 + v_2 && \qquad + d_2 \\
u_3 &= 2 + u_2 - v_2 + v_3 - c_1 + d_1 \\
z &= 3 + u_2 + 2v_1 + v_2 + 2v_3 - 2c_1 + 2d_1.
\end{aligned}
$$

Now c_1 becomes basic. To permit c_1 to increase beyond the value 1 requires u_1 and v_1 to be interchanged. The coefficient of c_1 in z would become $-2 - 2 \times (-1) = 0$ so that further decrease in z is not possible. Thus we interchange c_1 and u_1, obtaining

$$
\begin{aligned}
c_1 &= 1 - u_1 - u_2 + v_1 + v_2 && + d_1 \\
c_2 &= 2 - u_2 + v_2 && + d_2 \\
u_3 &= 1 + u_1 + 2u_2 - v_1 - 2v_2 + v_3 \\
z &= 1 + 2u_1 + 3u_2 - v_2 + 2v_3.
\end{aligned}
$$

This ends the first phase of the algorithm, and corresponds to an approximation to the solution with $Z = \{1, 2\}$. We now let v_2 become basic, exchanging in the usual way for u_3, since no increase beyond $v_2 = \frac{1}{2}$ is possible. This gives

$$
\begin{aligned}
c_1 &= \tfrac{3}{2} - \tfrac{1}{2}u_1 & -\tfrac{1}{2}u_3 + \tfrac{1}{2}v_1 + \tfrac{1}{2}v_3 + d_1 \\
c_2 &= \tfrac{5}{2} + \tfrac{1}{2}u_1 & -\tfrac{1}{2}u_3 - \tfrac{1}{2}v_1 + \tfrac{1}{2}v_3 \quad\;\; + d_2 \\
v_2 &= \tfrac{1}{2} + \tfrac{1}{2}u_1 + \; u_2 - \tfrac{1}{2}u_3 - \tfrac{1}{2}v_1 + \tfrac{1}{2}v_3 \\
z &= \tfrac{1}{2} + \tfrac{3}{2}u_1 + 2u_2 + \tfrac{1}{2}u_3 + \tfrac{1}{2}v_1 + \tfrac{3}{2}v_3.
\end{aligned}
$$

This is now the optimal solution, so that we have $Z = \{1, 3\}$, with $\mathbf{a} = (\tfrac{3}{2}, \tfrac{5}{2})^T$ and $\|\mathbf{r}\| = \tfrac{1}{2}$.

Barrodale and Roberts (1973) give numerical comparisons of this method with a number of others, including a method which solves the dual linear programming problem due to Robers and Ben-Israel (1969), and a direct descent method due to Usow (1967b) (we examine a more recent method of this type in the next section). The numerical results indicate that the method described here is the best of those tested.

Exercises

10. The L_1 norm is an example of a *polyhedral norm*: let B be an $N \times m$ matrix such that

(i) $C = \{\mathbf{r}: B\mathbf{r} \leqslant \mathbf{e}\}$ is bounded and has a non-empty interior
(ii) $\mathbf{r} \in C$ if and only if $-\mathbf{r} \in C$.

Then B defines the polyhedral norm on R^m

$$\|\mathbf{r}\|_B = \min\{\mu: B\mathbf{r} \leqslant \mu\mathbf{e}\}.$$

Determine the matrix B appropriate to the L_1 norm. If $\mathbf{r}(\mathbf{a}) = \mathbf{b} - A\mathbf{a}$, and if A has rank n, show that there exist $(n+1)$ components of the vector $B\mathbf{r}$ at which the norm μ is attained, for some $\mathbf{a}$ minimizing $\|\mathbf{r}\|_B$. Deduce the result of Theorem 6.2 in this case. (For an exchange method for the general linear polyhedral norm problem, see Anderson and Osborne, 1976.)

11. Show that any bounded, feasible linear programming problem can be expressed as a linear discrete L_1 problem.

12. By exercise 11, the linear discrete L_∞ problem may be solved through the solution of an L_1 problem. Show that (2.1) is precisely equivalent to the L_1 problem in R^{2m+1} defined by

$$
\mathbf{r} = \begin{bmatrix} A & \mathbf{e} \\ -A & \mathbf{e} \\ \mathbf{0}^T & 2m-1 \end{bmatrix} \begin{bmatrix} \mathbf{a} \\ h \end{bmatrix} - \begin{bmatrix} \mathbf{b} \\ -\mathbf{b} \\ \mu \end{bmatrix}
$$

where μ is a sufficiently large positive number. The value of h gives the minimum value of the L_∞ norm in (2.1).

13. Consider the method of Barrodale and Roberts (1973) as described here. If u_j is replaced in the basis by v_j, show that the coefficient of the nonbasic variable c_k

in the expression for z is changed from β_k to $\beta_k - 2\gamma_j$, where γ_j is the coefficient of v_j in the equation for u_j.

14.   If A has rank t, show that a basic feasible solution of (6.6) corresponds to an approximation at which Z contains (at least) t indices. What is the significance of Z containing more than t indices?

6.4 Direct descent methods

The linear programming problem (6.6) solved by the simplex method (or a variant) is a descent method for the L_1 problem. Unlike the corresponding L_∞ problem, there is no advantage to be gained in going to the dual problem, and thus defining an ascent method. However, alternative 'direct' descent methods analogous to those introduced in Chapter 2 can be of value. The method of Usow (1967b) has already been mentioned. Another approach is due to Claerbout and Muir (1973), and a more sophisticated variant of this to Bartels, Conn, and Sinclair (1978). We will describe the latter method.

   The essence of the approach is contained in the proof of necessity of the conditions of Theorem 6.1, for if a vector $\mathbf{c}$ satisfying (6.3) can be obtained, then $\mathbf{c}$ is a descent direction for $\|\mathbf{r}\|$. The particular descent direction used in the algorithm is conveniently defined in a direct manner as follows. We assume that we have an approximation $\mathbf{a} \in R^n$ to the solution of (6.1); then for any $\mathbf{c} \in R^n$, $\gamma > 0$,

$$\|\mathbf{r}(\mathbf{a} + \gamma\mathbf{c})\| = \sum_{i=1}^{m} \operatorname{sign}(r_i(\mathbf{a} + \gamma\mathbf{c}))(r_i(\mathbf{a}) - \gamma\alpha_i{}^T\mathbf{c}).$$

Let γ be sufficiently small that none of the nonzero components of $\mathbf{r}(\mathbf{a} + \gamma\mathbf{c})$ change sign. Then

$$\|\mathbf{r}(\mathbf{a} + \gamma\mathbf{c})\| = \sum_{i \in Z} \gamma|\alpha_i{}^T\mathbf{c}| + \sum_{i \notin Z} (|r_i(\mathbf{a})| - \gamma \operatorname{sign}(r_i(\mathbf{a}))\alpha_i{}^T\mathbf{c})$$

$$= \|\mathbf{r}(\mathbf{a})\| - \gamma\left(\mathbf{h}^T\mathbf{c} - \sum_{i \in Z} |\alpha_i{}^T\mathbf{c}|\right) \tag{6.7}$$

where $\mathbf{h} = \sum_{i \notin Z} \operatorname{sign}(r_i(\mathbf{a}))\alpha_i$. Now if $\alpha_i{}^T\mathbf{c} = 0$, $i \in Z$, we need only ensure that $\mathbf{h}^T\mathbf{c} > 0$ for $\mathbf{c}$ to be a descent direction. Let $Z = \{i_1, i_2, \ldots, i_r\}$, and assume that the rows of A corresponding to the indices in Z are linearly independent. Then if $r < n$, we can define the $n \times (n - r)$ matrix P by

$$P^T\alpha_i = \mathbf{0} \qquad i \in Z$$
$$P^TP = I$$

and can set $\mathbf{c} = PP^T\mathbf{h}$. Thus if $\mathbf{c} \neq \mathbf{0}$, $\mathbf{c}$ is a descent direction as required. Otherwise $P^T\mathbf{h} = \mathbf{0}$, and so

$$\mathbf{h} = \sum_{j=1}^{r} \lambda_j\alpha_{i_j} \tag{6.8}$$

for some numbers λ_j. (This relationship also holds when $r = n$, by assumption.) Using Theorem 6.1, it follows immediately from (6.8) that if $|\lambda_j| \leqslant 1$, $j = 1$, $2, \ldots, r$, then $\mathbf{a}$ solves (6.1). Otherwise if $|\lambda_{j_0}| > 1$ for some j_0, the norm can be reduced; we can see this directly as follows. Consider *any* vector $\mathbf{c} \in R^n$. Then

$$\mathbf{h}^T\mathbf{c} - \sum_{i \in Z} |\boldsymbol{\alpha}_i{}^T\mathbf{c}| = \sum_{j=1}^{r} \lambda_j \boldsymbol{\alpha}_{i_j}{}^T\mathbf{c} - \sum_{j=1}^{r} |\boldsymbol{\alpha}_{i_j}{}^T\mathbf{c}|$$

$$= \sum_{j=1}^{r} (\text{sign}\,(\boldsymbol{\alpha}_{i_j}{}^T\mathbf{c})\lambda_j - 1)|\boldsymbol{\alpha}_{i_j}{}^T\mathbf{c}|. \qquad (6.9)$$

Thus if $|\lambda_{j_0}| > 1$, this expression can be made positive, and so from (6.7) the norm can be reduced. Such a reduction can be achieved in the following way. Define P_0 as the $n \times (n - r + 1)$ matrix satisfying

$$P_0{}^T\boldsymbol{\alpha}_i = \mathbf{0} \qquad i \in \{Z - j_0\}$$
$$P_0{}^TP_0 = I,$$

and set

$$\mathbf{c} = \text{sign}\,(\lambda_{j_0})P_0P_0{}^T\boldsymbol{\alpha}_{i_{j_0}}.$$

Then

$$\boldsymbol{\alpha}_i{}^T\mathbf{c} = 0 \qquad i \in \{Z - j_0\}$$

and

$$\text{sign}\,(\boldsymbol{\alpha}_{i_{j_0}}{}^T\mathbf{c})\lambda_{j_0} = \text{sign}\,(\boldsymbol{\alpha}_{i_{j_0}}{}^T P_0 P_0{}^T\boldsymbol{\alpha}_{i_{j_0}})|\lambda_{j_0}|$$
$$= |\lambda_{j_0}|,$$

giving a positive value to (6.9), and showing that $\mathbf{c}$ is a descent direction.

Once either of the above processes has been used to define a direction $\mathbf{c}$, it remains to determine a suitable step length γ in this direction. A bound on the size of γ appears to be imposed by the requirement that there are no changes of sign in any of the components of $\mathbf{r}$ currently nonzero, and thus we may choose γ as the value when the first of these components becomes zero. Defining

$$\gamma_i = \frac{r_i(\mathbf{a})}{\boldsymbol{\alpha}_i{}^T\mathbf{c}} \qquad i \notin Z,$$

this gives

$$\gamma = \min_{i \notin Z} \{\gamma_i, \gamma_i > 0\}. \qquad (6.10)$$

If γ is marginally increased beyond this value, then $r_k(\mathbf{a} + \gamma\mathbf{c})$, $k \in K$ (say) will simultaneously change sign. In this case we will have (exercise 15)

$$\| \mathbf{r}(\mathbf{a} + \gamma \mathbf{c}) \| = \| \mathbf{r}(\mathbf{a}) \| - 2 \sum_{i \in K} |r_i(\mathbf{a})|$$

$$- \gamma (\mathbf{h}^T \mathbf{c} - 2 \sum_{i \in K} \text{sign} \, (r_i(\mathbf{a})) \boldsymbol{\alpha}_i{}^T \mathbf{c}). \qquad (6.11)$$

Provided that the coefficient of γ remains positive, then the value of the norm is still being decreased. Clearly this means that we may still be able to go on increasing γ beyond values at which other sets of components of $\mathbf{r}$ change sign. This process must eventually terminate when the appropriate coefficient of γ becomes negative (or else we could decrease $\| \mathbf{r} \|$ without limit); the current value of γ is then accepted.

When the rows of A corresponding to indices in Z are linearly dependent, then there is no unique vector $\boldsymbol{\lambda}$ satisfying (6.8). This degenerate situation may be resolved by further modification of the method, but we do not consider this here. In fact, although a theoretical resolution of this degenerate case is given by Bartels, Conn, and Sinclair (1978), they take the view that such difficulties are 'artifacts of the computer's limited word length'. In practice, the introduction of random perturbations has proved satisfactory when a degenerate situation is encountered. For other aspects of the method, including the efficient and stable treatment of the linear algebraic tasks involved, the reader is referred to the original paper. We illustrate the method by a simple example.

Example

$$A = \begin{bmatrix} 1 & 2 \\ -1 & 1 \\ 1 & 0 \\ 0 & -1 \end{bmatrix}, \qquad \mathbf{b} = \begin{bmatrix} 1 \\ 1 \\ 1 \\ 1 \end{bmatrix}.$$

Let $\mathbf{a} = (1, -1)^T$. Then $r_1 = 2, r_2 = 3, Z = \{3, 4\}$ and $\| \mathbf{r} \| = 5$. P is not defined, and

$$\mathbf{h} = \boldsymbol{\alpha}_1 + \boldsymbol{\alpha}_2 = (0, 3)^T.$$

Thus

$$\lambda_1 \begin{pmatrix} 1 \\ 0 \end{pmatrix} + \lambda_2 \begin{pmatrix} 0 \\ -1 \end{pmatrix} = \begin{pmatrix} 0 \\ 3 \end{pmatrix}$$

giving $\lambda_1 = 0, \lambda_2 = -3$. The conditions of Theorem 6.1 are not satisfied, and we take $j_0 = 4$. Then

$$P_0 = \begin{pmatrix} 0 \\ 1 \end{pmatrix}$$

and

$$c = (-1)\begin{pmatrix} 0 \\ 1 \end{pmatrix}(0, 1)\begin{pmatrix} 0 \\ -1 \end{pmatrix} = \begin{pmatrix} 0 \\ 1 \end{pmatrix}$$

Now $\gamma_1 = 2/2 = 1$, $\gamma_2 = 3/1 = 3$, and so $K = \{1\}$. Also if we define $\mathbf{g}$ by

$$\mathbf{g} = \mathbf{h} - 2 \operatorname{sign}(r_1(\mathbf{a}))\alpha_1 = (-2, -1)^T,$$

then $\mathbf{g}^T\mathbf{c} = -1$. Thus we take $\gamma = 1$, and the next approximation is $\mathbf{a} = (1, 0)^T$. Now $r_2 = 2$, $r_4 = 1$, $Z = \{1, 3\}$, and $\|\mathbf{r}\| = 3$. Again P is not defined, and

$$\mathbf{h} = \alpha_2 + \alpha_4 = (-1, 0)^T.$$

Thus

$$\lambda_1 \begin{pmatrix} 1 \\ 2 \end{pmatrix} + \lambda_2 \begin{pmatrix} 1 \\ 0 \end{pmatrix} = \begin{pmatrix} -1 \\ 0 \end{pmatrix}$$

giving $\lambda_1 = 0$, $\lambda_2 = -1$. We therefore have the solution.

Bartels, Conn, and Sinclair (1978) give some numerical comparisons of a number of examples solved by their algorithm and by the linear programming based method of Barrodale and Roberts (1973) described in the previous section. One obvious difficulty in making a comparison is that the former method requires a starting point to be provided, whereas the latter method does not. However, the *basic* step of each method is a move from one approximation with (generally) n indices in Z to another similar approximation. Thus, not surprisingly, the methods turn out to be very similar in practice, although they are not identical. This contrasts with the situation in the L_∞ problem, where the most efficient linear programming methods are ascent methods, and the direct descent methods seem to be at a disadvantage. Certainly, present indications are that for the L_∞ problem ascent methods should be used, while descent methods are more appropriate for the L_1 problem. This may be interpreted as reflecting the dual nature of the respective problems.

Exercises

15.   Obtain equation (6.11), and the corresponding expression if more than one set of components changes sign.

16.   Show that, in the nondegenerate case, the algorithm of this section converges in a finite number of steps.

17.   Show that, in the nondegenerate case, the values γ_i defined by (6.10) are such that $\gamma_i \neq \gamma_j$, $i \neq j$.

18.   Solve the example of this section by the method of the previous section, and vice versa (try different starting points).

7

Linear L_1 approximation of continuous functions

7.1 Introduction

The continuous analogue of the problem considered in the previous chapter is set in $C[X]$, X an N-dimensional continuum, and may be posed as follows. Given $f(\mathbf{x})$, $\phi_i(\mathbf{x})$, $i = 1, 2, \ldots, n$ all in $C[X]$,

$$\text{find } \mathbf{a} \in R^n \text{ to minimize } \| r(\mathbf{x}, \mathbf{a}) \| = \int_X |r(\mathbf{x}, \mathbf{a})| \, dx, \qquad (7.1)$$

where $r(\mathbf{x}, \mathbf{a}) = f(\mathbf{x}) - \sum\limits_{i=1}^{n} a_i \phi_i(\mathbf{x})$, $\mathbf{x} \in X$. A more natural setting for problems of this type is the space of integrable functions; in addition X need only be a compact, measurable subset of R^N. However, as in the corresponding L_p case, the present setting is appropriate for most frequently occurring problems. It is common to refer to (7.1) as 'approximation in the mean': for example, if $X = [a, b]$, then $\|f\|/(b-a)$ may be regarded as the mean or average value of $|f(x)|$ on X.

Following the established pattern, we begin by considering the characterization of solutions to (7.1). It will be assumed throughout this chapter that the set $\phi_i(\mathbf{x})$, $i = 1, 2, \ldots, n$ is linearly independent on X.

7.2 Characterization of solutions

Characterization results for (7.1) may be given in different forms, of which the one given here is perhaps the most convenient. An important role is played by the set $Z(\equiv Z(\mathbf{a}))$ of points $\mathbf{x} \in X$ for which $r(\mathbf{x}, \mathbf{a}) = 0$ (analogous to the set of *indices* Z of Chapter 6). We will also use $Z(\varepsilon)$ to denote the set

$$Z(\varepsilon) = \{\mathbf{x} : |r(\mathbf{x}, \mathbf{a})| \leqslant \varepsilon\},$$

which is defined for all $\varepsilon > 0$, $\theta(\mathbf{x}, \mathbf{a})$ to denote the (discontinuous) function satisfying

$$\theta(\mathbf{x}, \mathbf{a}) = \text{sign}(r(\mathbf{x}, \mathbf{a})) \qquad \mathbf{x} \in X, \qquad (7.2)$$

and $V(\mathbf{a})$ to denote the set

$$V(\mathbf{a}) = \{v(\mathbf{x}) : \max_{\mathbf{x} \in X} |v(\mathbf{x})| \leqslant 1,\ v(\mathbf{x}) = \theta(\mathbf{x}, \mathbf{a}),\ \mathbf{x} \notin Z\}.$$

Theorem 7.1 A vector $\mathbf{a} \in R^n$ solves (7.1) if and only if there exists $v(\mathbf{x}) \in V$ such that

$$\int_X v(\mathbf{x}) \phi_j(\mathbf{x}) \, dx = 0 \qquad j = 1, 2, \ldots, n. \qquad (7.3)$$

Proof 1   In the notation of Theorem 1.7, $V(r)$ is identical to the set V defined here, and so the result is immediate.

Proof 2   Let the conditions of the theorem be satisfied at $\mathbf{a}$. Then

$$\|r(\mathbf{x}, \mathbf{a})\| = \int_{X-Z} |r(\mathbf{x}, \mathbf{a})| \, dx$$

$$= \int_{X-Z} \theta(\mathbf{x}, \mathbf{a}) \left(f(\mathbf{x}) - \sum_{j=1}^{n} a_j \phi_j(\mathbf{x}) \right) dx$$

$$= \int_{X-Z} \theta(\mathbf{x}, \mathbf{a}) f(\mathbf{x}) \, dx + \int_Z v(\mathbf{x}) f(\mathbf{x}) \, dx. \qquad (7.4)$$

For any $\mathbf{d} \in R^n$

$$\|r(\mathbf{x}, \mathbf{d})\| \geqslant \int_{X-Z} \theta(\mathbf{x}, \mathbf{a}) r(\mathbf{x}, \mathbf{d}) \, dx + \int_X v(\mathbf{x}) r(\mathbf{x}, \mathbf{d}) \, dx$$

$$= \int_{X-Z} \theta(\mathbf{x}, \mathbf{a}) f(\mathbf{x}) \, dx + \int_Z v(\mathbf{x}) f(\mathbf{x}) \, dx$$

$$= \|r(\mathbf{x}, \mathbf{a})\|$$

using (7.4). Thus $\mathbf{a}$ solves (7.1).

Now let $\mathbf{a}$ be a solution to (7.1) and assume that the conditions of the theorem are not satisfied. Let

$$D = \left\{ \mathbf{d} \in R^n : d_j = \int_X v(\mathbf{x}) \phi_j(\mathbf{x}) \, dx, \quad v(\mathbf{x}) \in V, j = 1, 2, \ldots, n \right\}.$$

Then D is a closed, convex subset of R^n (exercise 1), with $\mathbf{0} \notin D$. Thus, by Theorem 1.5, there exists $\mathbf{c} \in R^n$, $\delta > 0$ such that

$$\mathbf{d}^T \mathbf{c} \geqslant \delta > 0 \qquad \text{for all } \mathbf{d} \in D$$

or

$$\sum_{j=1}^{n} c_j \int_X v(\mathbf{x}) \phi_j(\mathbf{x}) \, dx \geq \delta > 0 \qquad \text{for all } v(\mathbf{x}) \in V. \tag{7.5}$$

Now if $\gamma > 0$,

$$\|r(\mathbf{x}, \mathbf{a} + \gamma \mathbf{c})\| = \int_X \left| r(\mathbf{x}, \mathbf{a}) - \gamma \sum_{j=1}^{n} c_j \phi_j(\mathbf{x}) \right| dx$$

$$= \gamma \int_Z \left| \sum_{j=1}^{n} c_j \phi_j(\mathbf{x}) \right| dx$$

$$+ \int_{X-Z} \left| r(\mathbf{x}, \mathbf{a}) - \gamma \sum_{j=1}^{n} c_j \phi_j(\mathbf{x}) \right| dx.$$

Let

$$v(\mathbf{x}) = \begin{cases} -\operatorname{sign}\left(\sum_{j=1}^{n} c_j \phi_j(\mathbf{x}) \right) & \mathbf{x} \in Z \\ \theta(\mathbf{x}, \mathbf{a}) & \mathbf{x} \in X - Z. \end{cases}$$

Then $v(\mathbf{x}) \in V$, and so by (7.5)

$$\int_Z \left| \sum_{j=1}^{n} c_j \phi_j(\mathbf{x}) \right| dx \leq \int_{X-Z} \theta(\mathbf{x}, \mathbf{a}) \sum_{j=1}^{n} c_j \phi_j(\mathbf{x}) \, dx - \delta.$$

Further, let $M = \max\limits_{\mathbf{x} \in X} \left| \sum_{j=1}^{n} c_j \phi_j(\mathbf{x}) \right|$, and let $M\gamma = \varepsilon$.

Then

$$\|r(\mathbf{x}, \mathbf{a} + \gamma \mathbf{c})\| \leq -\gamma\delta + \gamma \int_{X-Z} \theta(\mathbf{x}, \mathbf{a}) \sum_{j=1}^{n} c_j \phi_j(\mathbf{x}) \, dx$$

$$+ \int_{X-Z} \left| r(\mathbf{x}, \mathbf{a}) - \gamma \sum_{j=1}^{n} c_j \phi_j(\mathbf{x}) \right| dx$$

$$= -\gamma\delta + \gamma \int_{X-Z(\varepsilon)} \theta(\mathbf{x}, \mathbf{a}) \sum_{j=1}^{n} c_j \phi_j(\mathbf{x}) \, dx$$

$$+ \gamma \int_{Z(\varepsilon)-Z} \theta(\mathbf{x}, \mathbf{a}) \sum_{j=1}^{n} c_j \phi_j(\mathbf{x}) \, dx + \int_{X-Z(\varepsilon)} |r(\mathbf{x}, \mathbf{a})| \, dx$$

$$- \gamma \int_{X-Z(\varepsilon)} \theta(\mathbf{x}, \mathbf{a}) \sum_{j=1}^{n} c_j \phi_j(\mathbf{x}) \, dx$$

$$+ \int_{Z(\varepsilon)-Z} \left| r(\mathbf{x}, \mathbf{a}) - \gamma \sum_{j=1}^{n} c_j \phi_j(\mathbf{x}) \right| dx$$

if $\gamma > 0$ is chosen sufficiently small, say $\gamma < G$, that for $0 < \gamma < G$

$$\text{sign}\left(r(\mathbf{x}, \mathbf{a}) - \gamma \sum_{j=1}^{n} c_j \phi_j(\mathbf{x}) \right) = \theta(\mathbf{x}, \mathbf{a}) \qquad \mathbf{x} \in X - Z(\varepsilon).$$

Thus if $\gamma < G$,

$$\| r(\mathbf{x}, \mathbf{a} + \gamma \mathbf{c}) \| \leqslant -\gamma \delta + \| r(\mathbf{x}, \mathbf{a}) \|$$

$$+ \int_{Z(\varepsilon) - Z} \left\{ \left| r(\mathbf{x}, \mathbf{a}) - \gamma \sum_{j=1}^{n} c_j \phi_j(\mathbf{x}) \right| \right.$$

$$\left. - |r(\mathbf{x}, \mathbf{a})| + \gamma \theta(\mathbf{x}, \mathbf{a}) \sum_{j=1}^{n} c_j \phi_j(\mathbf{x}) \right\} dx.$$

Now the integrand on the right hand side is bounded above in modulus by $4\varepsilon = 4\gamma M$. Thus

$$\| r(\mathbf{x}, \mathbf{a} + \gamma \mathbf{c}) \| \leqslant \| r(\mathbf{x}, \mathbf{a}) \| + \gamma(4M\mu(Z(\varepsilon) - Z) - \delta)$$

where the measure $\mu(Z(\varepsilon) - Z)$ tends to zero as $\varepsilon \to 0$ (or as $\gamma \to 0$). Thus if $\gamma < G$ is such that

$$4M\mu(Z(\varepsilon) - Z) < \delta$$

we have a contradiction that $\mathbf{a}$ solves (7.1) and the result is proved. □

Corollary Let $\mathbf{a} \in R^n$ be such that $\mu(Z) = 0$. Then $\mathbf{a}$ solves (7.1) if and only if

$$\int_X \theta(\mathbf{x}, \mathbf{a})\phi_j(\mathbf{x}) dx = 0 \qquad j = 1, 2, \ldots, n. \tag{7.6}$$

This corollary is an important one, for in general we would expect $\mu(Z) = 0$; for example this is the case if $r(\mathbf{x}, \mathbf{a})$ vanishes at only a finite number of points in X. More can be said about these interpolation points in the special case of the problem (7.1) when the set $\{\phi_i(\mathbf{x})\}$ is a Chebyshev set on X. We have the following analogue of Theorem 5.2, which may be proved in a similar manner (exercise 4).

Theorem 7.2 Let $X = [a, b]$ and let $\phi_i(x), i = 1, 2, \ldots, n$ form a Chebyshev set on X. Let $\mathbf{a}$ solve (7.1) with $\mu(Z) = 0$. Then $r(x, \mathbf{a})$ changes sign at least n times in $[a, b]$ if $\| r(x, \mathbf{a}) \| \neq 0$.

Exercises

1.   Show that the set D of Theorem 7.1 is convex.

2.   An alternative statement to that of Theorem 7.1 is the following: $\mathbf{a}$ solves (7.1) if and only if

$$\int_Z \left| \sum_{j=1}^{n} d_j \phi_j(\mathbf{x}) \right| dx \geqslant \left| \int_{X-Z} \theta(\mathbf{x}, \mathbf{a}) \sum_{j=1}^{n} d_j \phi_j(\mathbf{x}) dx \right|$$

for all $\mathbf{d} \in R^n$. Verify this.

3. For the special case when $X = [a, b]$, prove that $C[X]$ with the L_1 norm is not a strictly convex normed linear space.

4. Prove Theorem 7.2.

5. Let $X = [a, b]$, and let $\psi_1(x)$, $\psi_2(x)$, ... be a Markov set on $[a, b]$. For $k = 2, 3, \ldots$ let $\mathbf{c}^{(k)}$ solve (7.1) with $n = k - 1$, $\phi_i(x) = \psi_i(x)$, $i = 1, 2, \ldots, n$ and $f(x) = \psi_k(x)$.

Now redefine $\phi_i(x) = r(x, \mathbf{c}^{(i)})$, $i = 2, 3, \ldots$, with $\phi_1(x) = \psi_1(x)$. Show that $\phi_i(x)$, $i = 1, 2, \ldots$ is a Markov set on $[a, b]$. For given n and given $f(x) \in C[a, b]$, show that $\mathbf{a} \in R^n$ solves (7.1) in this case provided that

$$\text{sign}\, (r(x, \mathbf{a})) = \text{sign}\, (\phi_{n+1}(x)).$$

7.3 Uniqueness of the solution

The normed linear space occurring in this chapter is not strictly convex (exercise 3), and so additional assumptions are required to guarantee the uniqueness of the solution to (7.1). Unlike the situation in the corresponding discrete problem, a convenient condition for uniqueness can be given, and this turns out to be precisely the same condition as that required in the continuous Chebyshev approximation problem, viz. that the approximations be sought from a Haar subspace of $C[X]$. The following result was first proved by Jackson in 1921.

Theorem 7.3 Let $X = [a, b]$ and let $\phi_i(x)$, $i = 1, 2, \ldots, n$ form a Chebyshev set on $[a, b]$. Then the solution to (7.1) is unique.

Proof Let $\mathbf{a}$, $\mathbf{b}$ be two solutions to (7.1) with $\mathbf{a} \neq \mathbf{b}$. Then

$$\| r(x, \mathbf{a}) \| = \| r(x, \mathbf{b}) \| = h, \text{ say}$$

and by convexity $\| r(x, (\mathbf{a} + \mathbf{b})/2) \| = h$, also. It follows that

$$\int_a^b \left\{ \left| r\left(x, \frac{\mathbf{a} + \mathbf{b}}{2} \right) \right| - \tfrac{1}{2}|r(x, \mathbf{a})| - \tfrac{1}{2}|r(x, \mathbf{b})| \right\} dx = 0. \qquad (7.7)$$

Now, for any x,

$$\left| r\left(x, \frac{\mathbf{a} + \mathbf{b}}{2} \right) \right| \leqslant \tfrac{1}{2}|r(x, \mathbf{a})| + \tfrac{1}{2}|r(x, \mathbf{b})|$$

and so the integrand in (7.7) is nonpositive, and thus identically zero. It follows that any zero of $r(x, (\mathbf{a} + \mathbf{b})/2)$ is a zero of both $r(x, \mathbf{a})$ and $r(x, \mathbf{b})$ and thus $r(x, (\mathbf{a} + \mathbf{b})/2)$ has $k \leqslant n - 1$ distinct zeros in $[a, b]$, otherwise the Chebyshev set assumption contradicts $\mathbf{a} \neq \mathbf{b}$. However, Theorem 7.2 shows that in this case $r(x, (\mathbf{a} + \mathbf{b})/2)$ changes sign at least n times in $[a, b]$, and so again we have a contradiction. The original assumption that $\mathbf{a} \neq \mathbf{b}$ must therefore be impossible, and the result is proved. $\qquad \square$

In view of this result, and Theorem 3.6, it might be supposed that the Chebyshev set assumption would also be sufficient for *strong uniqueness* of the solution to (7.1). This is, however, not the case, as the following example shows.

Example $X = [0, 1]$, $n = 1$, $f(x) = x^2$, $\phi_1(x) = 1$.

$$\|r(x, a)\| = \int_0^1 |x^2 - a|\, dx$$

$$= \int_0^{\sqrt{a}} (a - x^2)\, dx + \int_{\sqrt{a}}^1 (x^2 - a)\, dx \qquad \text{if } 0 \leqslant \sqrt{a} \leqslant 1$$

$$= \tfrac{4}{3}|a|^{3/2} - a + \tfrac{1}{3}.$$

This expression is minimized by the choice $a = \tfrac{1}{4}$, with $\|r\| = \tfrac{1}{4}$ and this is clearly the minimum value of the norm. Now for any c such that $\sqrt{c} \in [\tfrac{1}{2}, 1]$,

$$\frac{\|r(x, c)\| - \tfrac{1}{4}}{|c - \tfrac{1}{4}|} = \frac{\tfrac{4}{3}c^{3/2} - c + 1/12}{c - \tfrac{1}{4}} \to 0 \text{ as } \sqrt{c} \to \tfrac{1}{2}+.$$

Thus $a = \tfrac{1}{4}$ is not a strongly unique solution to (7.1).

The method used for calculating the solution in this example illustrates an essential difference which exists between this class of problems and the class (3.1): it is possible in the Chebyshev set case for (7.1) to be equivalent to the minimization of a function differentiable at the solution. This precludes strong uniqueness.

We conclude this section by considering briefly the situation when the continuity requirement in (7.1) is relaxed, so that $C[X]$ is replaced by an appropriate space $L^1[X]$ of integrable functions. This case is considered in some detail by Kripke and Rivlin (1965), who show, for example, that the important corollary to Theorem 7.1 still holds. However, Theorem 7.3 is no longer valid, as the following example from Carroll and McLaughlin (1973) illustrates.

Example $X = [-1, 3]$, $n = 2$, $\phi_1(x) = 1$, $\phi_2(x) = x$,

$$f(x) = \begin{cases} 0 & -1 \leqslant x \leqslant 2 \\ -1 & 2 < x \leqslant 3. \end{cases}$$

Let $a_1 = 0$, $a_2 = t$ where $-\tfrac{1}{3} \leqslant t \leqslant 0$. Then

$$\int_{-1}^3 1 \operatorname{sign}(f(x) - tx)\, dx = 0$$

$$\int_{-1}^3 x \operatorname{sign}(f(x) - tx)\, dx = 0,$$

so that tx is a best approximation for all t such that $-\tfrac{1}{3} \leqslant t \leqslant 0$. The details of this example are left as an exercise.

Now let $\phi_1(x), \phi_2(x), \ldots, \phi_n(x)$ be elements of $L^1[X]$ such that the best approximation is unique for *all* $f(x) \in L^1[X]$. Then Wulbert (1971) has shown that for any $f(x) \in L^1[X]$, the solution to (7.1) is *strongly unique*. This result may be regarded as the analogue of Theorem 3.6.

Exercises

6. Complete the details of the last example above.

7. Let $X = [0, 1]$, $n = 2$, $\phi_1(x) = 1$, $\phi_2(x) = x$,

$$f(x) = \begin{cases} x & 0 \leqslant x \leqslant \frac{1}{2} \\ 1 - x & \frac{1}{2} \leqslant x \leqslant 1. \end{cases}$$

Obtain analytic expressions for $\|r(x, \mathbf{a})\|$ in each of 7 different regions of the (a_1, a_2) plane, and hence solve the problem (7.1) in this case. Is the solution strongly unique? (Marti, 1975.)

7.4 The construction of best approximations

The problem of the actual calculation of best L_1 approximations to continuous functions is of theoretical, rather than practical interest, and is not one which has received a great deal of attention, at least in any generality. We will only concern ourselves with the particular case when X is the interval $[a, b]$, for this enables some special aspects of the theory to be utilized. We have seen that for the discrete L_1 problem, there is a close relationship with an appropriate interpolation problem, and Theorem 7.2 suggests that a similar connection may be drawn in the continuous case.

Suppose that we have a set of points $a = x_0 < x_1 < x_2 < \ldots < x_n < x_{n+1} = b$ and a sign function $s(x)$ satisfying $|s(x)| = 1$, $x \in (x_i, x_{i+1})$, which can change value at each point x_i, $i = 1, 2, \ldots, n$ and at no other point in $[a, b]$. Suppose further that

$$\int_a^b s(x)\phi_j(x)dx = 0 \qquad j = 1, 2, \ldots, n. \tag{7.8}$$

Now let $\mathbf{a} \in R^n$ satisfy the interpolation conditions

$$\sum_{j=1}^n a_j \phi_j(x_i) = f(x_i) \qquad i = 1, 2, \ldots, n. \tag{7.9}$$

Then if $s(x) = \theta(x, \mathbf{a})$, it follows from (7.8) and the Corollary to Theorem 7.1 that $\mathbf{a}$ solves (7.1). Of course, there is no guarantee that a solution to (7.9) will be such that $s(x) = \theta(x, \mathbf{a})$, and so even if an appropriate $s(x)$ satisfying (7.8) can be obtained, this need not lead directly to a solution of (7.1). The existence of a sign function satisfying (7.8) is shown by Hobby and Rice (1965). When $\phi_j(x)$, $j = 1, 2, \ldots, n$, form a Chebyshev set on $[a, b]$, they also show that the sign function is unique, and has exactly n sign changes. We can illustrate this for the special case of degree $n-1$ polynomial approximation in the interval $[-1, 1]$, when $s(x)$ must satisfy

$$\int_{-1}^{1} x^k s(x)\,dx = 0 \qquad k = 0, 1, \ldots, n-1. \tag{7.10}$$

We will show that an appropriate $s(x)$ is furnished by taking

$$s(x) = \text{sign}\,(U_n(x))$$

where $U_n(x)$ is the Chebyshev polynomial of the second kind defined by

$$U_n(x) = \frac{\sin(n+1)\theta}{\sin\theta},$$

where $x = \cos\theta$. Now

$$\int_{-1}^{1} x^k \,\text{sign}\,(U_n(x))\,dx = \int_0^{\pi} (\cos\theta)^k \,\text{sign}\left(\frac{\sin(n+1)\theta}{\sin\theta}\right) \sin\theta\,d\theta$$

$$= \int_0^{\pi} (\cos\theta)^k \sin\theta\,\text{sign}\,(\sin(n+1)\theta)\,d\theta$$

$$= \tfrac{1}{2}\int_{-\pi}^{\pi} (\cos\theta)^k \sin\theta\,\text{sign}\,(\sin(n+1)\theta)\,d\theta.$$

Also

$$(\cos\theta)^k \sin\theta = \frac{(e^{i\theta}+e^{-i\theta})^k\,(e^{i\theta}-e^{-i\theta})}{2^{k+1}i}$$

$$= \sum_{|p|\leqslant k+1} \alpha_p e^{ip\theta}$$

for some constants α_p. Thus the required result (7.10) will hold provided that

$$I_p = \int_{-\pi}^{\pi} e^{ip\theta}\,\text{sign}\,(\sin(n+1)\theta)\,d\theta = 0 \qquad p = 0, \pm 1, \ldots, \pm n.$$

Putting $\theta = \phi + \pi/(n+1)$, we have

$$I_p = \int_{-\pi-\pi/(n+1)}^{\pi-\pi/(n+1)} e^{ip(\phi+\pi/(n+1))}\,\text{sign}\,(\sin(n+1)(\phi+\pi/(n+1)))\,d\phi$$

$$= -e^{ip\pi/(n+1)}\int_{-\pi-\pi(n+1)}^{\pi-\pi/(n+1)} e^{ip\phi}\,\text{sign}\,(\sin(n+1)\phi)\,d\phi$$

$$= -e^{ip\pi/(n+1)}I_p$$

since the integrand is periodic, period 2π. Since $-e^{ip\pi/(n+1)} \neq 1, |p| \leqslant n$, it follows that $I_p = 0, |p| \leqslant n$, and so we have shown that

$$\int_{-1}^{1} x^k \,\text{sign}\,(U_n(x))\,dx = 0 \qquad k = 0, 1, \ldots, n-1. \tag{7.11}$$

Now the zeros of $U_n(x)$ in $[-1, 1]$ are those of $\sin(n+1)\theta/\sin\theta$ in $[0, \pi]$. Since $\sin\theta = 0$ at $\theta = 0$, π at which $U_n(x) \neq 0$, the zeros are given by

$$(n+1)\theta_k = k\pi \qquad k = 1, 2, \ldots, n,$$

or

$$x_k = \cos\theta_k = \cos\left(\frac{k\pi}{n+1}\right) \qquad k = 1, 2, \ldots, n. \quad (7.12)$$

Thus, if $\mathbf{a} \in R^n$ is such that

$$p(x_k) \equiv \sum_{i=1}^{n} a_i x_k^{i-1} = f(x_k) \qquad k = 1, 2, \ldots, n, \quad (7.13)$$

with no other points of interpolation in $[-1, 1]$,

$$\text{sign}(p(x) - f(x)) = \pm\text{sign}(U_n(x)) \qquad -1 \leqslant x \leqslant 1$$

and so by (7.11)

$$\int_{-1}^{1} x^k \text{sign}(p(x) - f(x)) dx = 0 \qquad k = 0, 1, \ldots, n-1.$$

In other words, if $p(x) - f(x)$ is zero *only* at the points $x_k, k = 1, 2, \ldots, n$ defined by (7.12), then $p(x)$ is the best L_1 approximation by a degree $(n-1)$ polynomial to $f(x)$ in $[-1, 1]$. This result is remarkable in that the points x_k are fixed, *independent of $f(x)$*.

In fact, the role played by the Chebyshev polynomial of the second kind in L_1 approximation is analogous to that played by the Chebyshev polynomials of the first kind in L_∞ approximation, as exemplified by Theorem 5.7. We have the following parallel result for degree n approximation.

Theorem 7.4 Of all monic polynomials of degree n, the Chebyshev polynomial $2^{-n}U_n(x)$ is the best L_1 approximation to zero in $[-1, 1]$.

Proof From the definition, $U_0(1) = 1$, $U_1(x) = 2x$ and so from the recurrence relation (exercise 23(ii) of Chapter 5), it follows that the coefficient of x^n in U_n is 2^n. Thus $2^{-n}U_n$ is indeed monic, and so $2^{-n}U_n(x) - x^n$ is a polynomial of degree $(n-1)$. Now equations (7.11) give

$$\int_{-1}^{1} x^k \text{sign}(2^{-n}U_n(x) - x^n + x^n) dx = 0 \qquad k = 0, 1, \ldots, n-1$$

so that by the corollary to Theorem 7.1, $2^{-n}U_n(x) - x^n$ is the best degree $n-1$ polynomial approximation to $f(x) = -x^n$ in the L_1 norm. The required result follows immediately. $\square$

Recalling the role played by Legendre polynomials in L_2 approximation, we have now seen that the problem of finding the monic polynomial of degree n which is the best L_p approximation to zero in $[-1, 1]$ is effectively solved by finding polynomials orthogonal with respect to the weight function $(1-x^2)^{1/p-1/2}$ in $[-1, 1]$ for the cases $p = 1, 2, \infty$. It is tempting therefore to look for this result to extend to the remaining values of p. However, Burgoyne (1967) shows by a counterexample that this cannot be done.

For the treatment of more general L_1 approximation problems, it is essential to make the assumption that $\mu(Z) = 0$ at the solution, so that the problem reduces to that of finding $\mathbf{a} \in R^n$ to satisfy

$$\int_a^b \theta(x, \mathbf{a})\phi_j(x)\mathrm{d}x = 0 \qquad j = 1, 2, \ldots, n. \tag{7.14}$$

This is just a system of n nonlinear equations in n unknowns, although of a rather unsual and inconvenient form, so that a direct approach to their solution is unlikely to be fruitful. However if it is known that at the solution, $r(x, \mathbf{a})$ changes sign exactly m times in $[a, b]$, then (7.14) can be effectively replaced by the equations

$$\sum_{i=1}^{m+1} (-1)^i \int_{x_{i-1}}^{x_i} \phi_j(x)\mathrm{d}x = 0 \qquad j = 1, 2, \ldots, n,$$

$$\sum_{j=1}^n a_j\phi_j(x_i) - f(x_i) = 0 \qquad i = 1, 2, \ldots, m,$$

where $x_0 = a$ and $x_{m+1} = b$. This is a system of $(n+m)$ nonlinear equations in the $(n+m)$ unknowns $\mathbf{a}, x_1, x_2, \ldots, x_m$, of a more tractable form than (7.14). The value of m, and good approximations to the unknowns, may be obtained by solving a *discretization* of the problem (7.1) (see the following section), when the methods of the previous chapter are available. A method of this type, which uses Newton's method to solve the nonlinear system, is given by Glashoff and Schultz (1979). Notice that this is the analogue of the method for the continuous L_∞ problem given at the start of Section 3.6.

Another possible way of solving (7.1) is by means of a descent method: if $\mathbf{a}$ is an approximation to the solution not satisfying (7.14), then a reduction in the value of the norm may be achieved by a sufficiently small step in any direction $\mathbf{c}$ satisfying (7.5). In this case we may write

$$\sum_{j=1}^n c_j \int_a^b \theta(x, \mathbf{a})\phi_j(x)\mathrm{d}x \geqslant \delta > 0 \tag{7.15}$$

if it is assumed that $\mu(Z) = 0$ at the current approximation. Thus we may choose, for example,

$$c_j = \int_a^b \theta(x, \mathbf{a})\phi_j(x)\mathrm{d}x \qquad j = 1, 2, \ldots, n,$$

which only fails to define a descent direction if $\mathbf{a}$ is a solution.

An algorithm of descent type is given by Usow (1967a), who gives an analysis applicable for problems with Chebyshev sets. His numerical results are for polynomials, and then a good first approximation may be obtained by satisfying (7.13). Indeed, as might be expected, this often gives the solution immediately. The method, however, does not *always* work and Marti (1975) gives an example of the approximation to a continuous function by a polynomial of degree one for which Usow's method converges to a point which is not a solution. An alternative descent method is given by Marti (1975) which has guaranteed convergence to a solution of (7.1) when the functions $\phi_1(x), \phi_2(x), \ldots, \phi_n(x)$ form a Markov set in $[a, b]$.

Exercises

8. Let $X = [-1, 1]$ and let $\phi_i(x) = x^{i-1}$, $i = 1, 2, \ldots, n$. If $\{\phi_1(x), \phi_2(x), \ldots, \phi_n(x), f(x)\}$ forms a Chebyshev set on X, show that the solution to (7.1) is given by the vector of coefficients of the polynomial of degree $(n-1)$ which interpolates $f(x)$ at the points $x_i = \cos(i\pi/n)$, $i = 1, 2, \ldots, n$.

9. Prove that

$$\int_0^\pi (\sin kx)\,\text{sign}(\sin(n+1)x)dx = 0 \qquad k = 1, 2, \ldots, n.$$

10. Let $X = [0, \pi]$, $f(x) = x$, $\phi_j(x) = \sin jx$, $j = 1, 2, \ldots, n$. Prove that (7.1) is solved by $\mathbf{a} \in R^n$ which satisfies

$$\sum_{j=1}^n a_j \sin j x_i = x_i \qquad j = 1, 2, \ldots, n$$

where $x_i = i\pi/(n+1)$, $i = 1, 2, \ldots, n$. (Hint: use the fact that $\phi_j(x)$, $j = 1, 2, \ldots, n$, is a Chebyshev set on $(0, \pi)$, and also exercise 9.)

Deduce that the minimum value of the norm is $\pi^2/2(n+1)$.

7.5 Discretization

For best Chebyshev approximations on a continuum X, great importance is attached to corresponding discrete problems obtained by replacing X by a discrete set of points in X. In the L_1 case, the relationship between these problems is not as close. However, we do have an analogue of the result (Theorem 3.7) concerning discrete sets which more and more 'fill out' X. We will confine the problem to the case where $X = [a, b]$, and consider approximations on discrete sets $Y = \{x_1, x_2, \ldots, x_m\}$ chosen so that $x_i \in [y_{i-1}, y_i]$, $i = 1, 2, \ldots, m$ where $a \leqslant y_0 < y_1 < \ldots < y_m \leqslant b$. Let $h = \max_{1 \leqslant i \leqslant m} h_i$ where $h_i = y_i - y_{i-1}$, $i = 1, 2, \ldots, m$, let Ω be as defined in Theorem 3.7, and define ω by

$$\omega(f, h) = \sup_{|x-y| \leqslant h} |f(x) - f(y)|.$$

Lemma 7.1 Let $g(x) \in C[a, b]$. Then

$$\left| \int_a^b g(x)dx - \sum_{i=1}^m g(x_i)h_i \right| \leqslant (b-a)\omega(g, h).$$

Proof

$$
\begin{aligned}
\left| \int_a^b g(x)dx - \sum_{i=1}^m g(x_i)h_i \right| &= \left| \sum_{i=1}^m \int_{y_{i-1}}^{y_i} [g(x) - g(x_i)]dx \right| \\
&\leqslant \sum_{i=1}^m \int_{y_{i-1}}^{y_i} |g(x) - g(x_i)|dx \\
&\leqslant \omega(g, h) \sum_{i=1}^m h_i = (b-a)\omega(g, h). \qquad \square
\end{aligned}
$$

Now let us define $M = \min_{\mathbf{a}} \| r(x, \mathbf{a}) \|$, and

$$M(h) = \min_{\mathbf{a}} \sum_{i=1}^m |r(x_i, \mathbf{a})|h_i = \sum_{i=1}^m |r(x_i, \mathbf{a}(h))|h_i.$$

Then we have the following analogue of Theorem 3.7, due essentially to Motzkin and Walsh (1956); see also Usow (1967b).

Theorem 7.5 Let $h \to 0$. Then
    (i) $M(h) \to M$
    (ii) $\| r(x, \mathbf{a}(h)) \| \to M$
    (iii) If $\mathbf{a}^*$ is the unique solution to (7.1), $\mathbf{a}(h) \to \mathbf{a}^*$.

Proof
By continuity, $\omega(f, h)$ and $\Omega(h)$ tend to zero with h. Let h be such that $\omega(f, h) \leqslant \| f \|/(b-a)$, $\Omega(h) < \delta$, where $\delta > 0$ satisfies

$$\left\| \sum_{j=1}^n c_j \phi_j(x) \right\| \geqslant \delta \left| \sum_{j=1}^n c_j \right|$$

for all $\mathbf{c} \in R^n$. Now

$$
\begin{aligned}
M &\leqslant \int_a^b |r(x, \mathbf{a}(h)|dx \\
&\leqslant M(h) + (b-a)\omega(|r(x, \mathbf{a}(h))|, h) \text{ by Lemma 7.1} \\
&\leqslant M(h) + (b-a)\omega(r(x, \mathbf{a}(h)), h) \\
&\leqslant M(h) + (b-a)\left| \sum_{j=1}^n a_j(h) \right| \Omega(h). \qquad (7.16)
\end{aligned}
$$

Further,

$$\left| \sum_{j=1}^{n} a_j(h) \right| \le \frac{1}{\delta} \left\| \sum_{j=1}^{n} a_j(h)\phi_j(x) \right\|$$

$$\le \frac{2}{\delta} \sum_{i=1}^{m} \left| \sum_{j=1}^{n} a_j(h)\phi_j(x_i) \right| h_i \qquad \text{by exercise 12,}$$

$$= \frac{2}{\delta} \sum_{i=1}^{m} \left| \sum_{j=1}^{n} a_j(h)\phi_j(x_i) - f(x_i) + f(x_i) \right| h_i$$

$$\le \frac{4}{\delta} \sum_{i=1}^{m} |f(x_i)| h_i$$

$$\le \frac{4}{\delta}(\|f\| + (b-a)\,\omega(f,h)) \le \frac{8}{\delta}\|f\|. \tag{7.17}$$

Thus

$$M \le M(h) + \frac{8(b-a)}{\delta} \|f\|\Omega(h); \tag{7.18}$$

also if $\mathbf{a}^*$ is *any* solution to (7.1),

$$M = \sum_{i=1}^{m} |r(x_i, \mathbf{a}^*)| h_i + E(h)$$

where $E(h) \to 0$ as $h \to 0$. Thus

$$M \ge M(h) + E(h) \tag{7.19}$$

and so (7.18) and (7.19) give (i). The inequalities to (7.16) then give (ii), since the inequalities to (7.17) show that $\{\mathbf{a}(h)\}$ is bounded. Finally, the limit points, by (ii), must solve (7.1), and the result (iii) follows. $\qquad \square$

Theorem 7.5 shows that a good approximate value of the minimum norm in (7.1) may be obtained by using one of the methods of Chapter 6. In order to make such a method efficient, we would like (as in the L_∞ case) to be rather more selective in the choice of the discrete set at each stage. However, the way to proceed is no longer straightforward. The natural analogue of the L_∞ methods would be to construct a sequence of discrete problems whose solutions are defined by n interpolation points from the discrete set, and such that these points 'converge' to the appropriate set of interpolation points in $[a, b]$ which essentially characterize the solution to the continuous problem. Any method of this type would be unlikely to avoid a Chebyshev set assumption. In addition, there is no obvious way to position the new points at each step in order to ensure convergence. While such fundamental difficulties remain unresolved, the problem of providing a good general method of this type for efficiently solving (7.1) remains open.

144

Exercises

11. Let $f(x) \in C[a, b]$ be such that $\|f\| > 0$, and let h be such that $\omega(f, h) < \|f\|/(b-a)$. Prove that

$$\|f\| \leqslant h \sum_{i=1}^{m} |f(x_i)| \left[1 + \frac{(b-a)\omega(f, h)}{\|f\| - (b-a)\omega(f, h)} \right].$$

12. Let $\Omega(h) < \delta$. Use the method of the previous exercise to prove that

$$\left\| \sum_{j=1}^{n} a_j \phi_j(x) \right\| \leqslant h \sum_{i=1}^{m} \left| \sum_{j=1}^{n} a_j \phi_j(x_i) \right| \left[1 + \frac{(b-a)\Omega(h)}{\delta - (b-a)\Omega(h)} \right]$$

for any $\mathbf{a} \in R^n$.

13. Establish the analogue of Theorem 7.5 for the L_p norms, $1 < p < \infty$.

8

Piecewise polynomial approximating functions

8.1 Introduction

This chapter differs from the others in this book in that we do not consider here a particular class of approximation problems; rather we introduce a particular class of *approximating functions*, and examine some of the important properties of members of this class. Most of the approximating functions introduced so far are such that their behaviour in a small region determines their behaviour everywhere in the region of interest. The fitting of such functions assumes that in those parts of the approximation over which we have no immediate control, there will be a matching uniformity of behaviour in the given data in order that a reasonable fit be obtained over the whole region. However, in many physical problems, the behaviour of the data, or the solution to which an approximation is required, may in one region be considerably different, and in fact totally unrelated, to its behaviour in another. The fitting of a mathematical function which has uniform smoothness properties, for example, in the whole of the region of interest may therefore be totally inappropriate. Such problems arise in the modelling of a number of physical quantities, such as coast lines or roads, or indeed in solving many differential equations. For example, suppose that such data has the form shown in Figure 11. The fitting of a curve to this in the interval $[a, b]$ may be achieved by taking the approximation to be a polynomial of degree, say, 3 or 4. However, the particular form of the given curve suggests that much better results would be achieved by considering two subregions $[a, y]$ and $[y, b]$ inside which *different* approximations would be appropriate. For example, a straight line will give a good fit in $[a, y]$, and a quadratic polynomial in $[y, b]$. Such an approximation will be defined by at most 4 parameters, if continuity is enforced; this is the same as a degree 3 polynomial, but will clearly result in a much better fit.

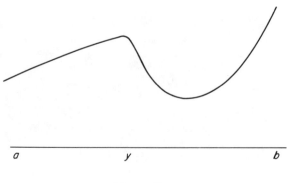

Figure 11

The idea behind this rather simple example, then, is to motivate the use of *piecewise* approximating functions. Essentially, the idea is to divide up the region in which an approximation is required into several subregions, and fit (possibly) different functions in each. Any mathematical functions may be used in each subregion, but we will restrict consideration to *piecewise polynomials* (possibly defined in more than one variable) as they are easy to use, and any lack of flexibility may be absorbed in the splitting of the region. It would clearly be unreasonable to fit these curves entirely independently of each other: if continuity of certain derivatives is forced at the edges of the subregions, then we can ensure that over the whole of the original region, a physically smooth curve will be produced. Of great importance in this process will be the way in which the region is subdivided. In one dimension, this corresponds to placement of points (called *knots* in this case), and unlike the simple example above, ideal positions will not generally be known beforehand. The next best thing would seem to be to have these knots as variables of the problem, to be shifted into the optimal positions as the computation proceeds. However, this turns out to be a very difficult nonlinear problem, and so the knot positions are usually fixed in advance.

Generally, if the subdivisions of the original region are specified *a priori*, then any appropriate piecewise polynomial can be regarded as an element of a finite dimensional linear space, the dimension corresponding to the number of degrees of freedom of the function. If an appropriate set of basis functions $\phi_1(\mathbf{x}), \ldots, \phi_n(\mathbf{x})$ is obtained, then the piecewise polynomial may be represented as $\sum_{i=1}^{n} a_i \phi_i(\mathbf{x})$ and the analysis and methods of the previous chapters are available for the construction of such functions. Because the nature of the problem often forces the value of n to be large, computational considerations dictate that approximations be obtained in a fairly simple manner, and the determination of suitable vectors $\mathbf{a}$ by interpolation, for example, is important. Piecewise polynomials are also frequently used to approximate functions defined *implicitly*, for example by differential equations. The most common methods of solution of such problems have important consequences in the way in which the approximating piecewise polynomials are represented, as we shall see later.

We now introduce two important classes of piecewise polynomial space. The analysis is carried out for functions of one variable, with an indication given of the higher dimensional extensions.

8.2 The Hermite space $H^{(m)}(\Delta)$

Let X be the interval $[a, b]$ of the real line, and let $C^m[X]$ denote the space of m times continuously differentiable functions defined on X. Let Δ denote the discrete set of N points in (a, b) defined by

$$a = x_0 < x_1 < x_2 \ldots < x_N < x_{N+1} = b.$$

Definition 8.1 The Hermite space $H^{(m)}(\Delta)$ is given by
$H^m(\Delta) = \{g(x) \in C^{m-1}[X]; g(x) \text{ is a polynomial of degree } 2m - 1 \text{ on each interval}$
$[x_i, x_{i+1}], i = 0, 1, \ldots, N, \text{ defined by } \Delta\}.$

This class of approximating functions thus consists of piecewise polynomials of degree $(2m - 1)$ which have $(m - 1)$ continuous derivatives at the grid points or knots $x_i, i = 1, 2, \ldots, N$. Since there are a total of $(N + 1)$ different polynomials and m continuity conditions at each knot, the number of degrees of freedom of each element of $H^{(m)}(\Delta)$ is $2m(N + 1) - Nm = m(N + 2)$, so that the linear space $H^{(m)}(\Delta)$ has dimension $m(N + 2)$. Thus, a particular element of $H^{(m)}(\Delta)$ will in general be defined by making it satisfy $m(N + 2)$ linear conditions. Let D^k denote the differential operator of order k with respect to x.

Theorem 8.1 For any prescribed set of values f_i^k, $i = 0, 1, \ldots, N + 1$, $k = 0, 1, \ldots, m - 1$, there is a unique element $g(x) \in H^{(m)}(\Delta)$ such that

$$D^k g(x_i) = f_i^k \qquad i = 0, 1, \ldots, N + 1; k = 0, 1, \ldots, m - 1. \tag{8.1}$$

Proof In each interval $[x_i, x_{i+1}]$, there are $2m$ conditions which uniquely determine a polynomial of degree $2m - 1$ (the details of this are left as an exercise). It is obvious that the appropriate continuity requirements are satisfied. □

If the numbers f_i^k are regarded as function values and derivatives of order k at the points x_i, $i = 0, 1, \ldots, N + 1$, $k = 0, 1, \ldots, m - 1$, then $g(x)$ and its derivatives interpolate these quantities at each point. We therefore have a special case of Hermite interpolation, and since any element of $H^{(m)}(\Delta)$ can be defined in this way, this explains the naming of the space. Theorem 8.1 also motivates the provision of a basis for $H^{(m)}(\Delta)$: for $i = 0, 1, \ldots, N + 1$, and $k = 0, 1, \ldots, m - 1$, define functions $\phi_{ik}(x)$ by

$$D^l \phi_{ik}(x_j) = \delta_{ij}\delta_{lk} \qquad l = 0, 1, \ldots, m - 1; j = 0, 1, \ldots, N + 1 \tag{8.2}$$

where δ_{ij} is the Kronecker delta. Then $\phi_{ik}(x) \in H^{(m)}(\Delta)$ by Theorem 8.1, and for any $g(x)$ defined by this theorem, we can write

$$g(x) = \sum_{i=0}^{N+1} \sum_{k=0}^{m-1} f_i^k \phi_{ik}(x). \tag{8.3}$$

For given k, i

$$D^l \phi_{ik}(x_j) = 0 \quad l = 0, 1, \ldots, m-1; \quad j = 0, 1, \ldots, i-1, i+1, \ldots, N+1$$

and so it follows that $\phi_{ik}(x)$ *vanishes identically* outside $[x_{i-1}, x_{i+1}]$: $\phi_{ik}(x)$ is said to have *compact support*, with its support contained in $[x_{i-1}, x_{i+1}]$.

Example $m = 1$

The elements of $H^{(1)}(\Delta)$ are *piecewise linear functions*. Writing $f_i^0 = f_i$, $\phi_{i0}(x) = l_i(x)$, $i = 0, 1, \ldots, N+1$, we have

$$g(x) = \sum_{i=0}^{N+1} f_i l_i(x), \tag{8.4}$$

where $l_i(x_j) = \delta_{ij}$, $i, j = 0, 1, \ldots, N+1$.

Letting $h_i = x_{i+1} - x_i$, the basis functions $l_i(x)$ are explicitly given by (see also Figure 12)

$$l_0(x) = \begin{cases} (x_1 - x)/h_0 & x_0 \leqslant x \leqslant x_1 \\ 0 & x_1 \leqslant x \leqslant x_{N+1} \end{cases}$$

$$l_i(x) = \begin{cases} 0 & x_0 \leqslant x \leqslant x_{i-1} \\ (x - x_{i-1})/h_{i-1} & x_{i-1} \leqslant x \leqslant x_i \\ (x_{i+1} - x)/h_i & x_i \leqslant x \leqslant x_{i+1} \\ 0 & x_{i+1} \leqslant x \leqslant x_{N+1} \end{cases}$$

$$l_{N+1}(x) = \begin{cases} 0 & x_0 \leqslant x \leqslant x_N \\ (x - x_N)/h_N & x_N \leqslant x \leqslant x_{N+1}. \end{cases}$$

The basis functions $l_i(x)$ are often referred to as 'hat functions' or 'roof functions'.

If the partition of the interval $[a, b]$ is *uniform* so that $h_i = h$, $i = 0, 1, \ldots, N$, then we can express all the functions $l_i(x)$ in terms of one 'standard' basis function $l(x)$. If we define

$$l(x) = \begin{cases} 0 & x \leqslant -1 \\ 1 + x & -1 \leqslant x \leqslant 0 \\ 1 - x & 0 \leqslant x \leqslant 1 \\ 0 & 1 \leqslant x \end{cases}$$

then it may be shown that

$$l_i(x) = l\left(\frac{x - x_0}{h} - i\right) \quad i = 0, 1, \ldots, N+1.$$

The verification of this is left as an exercise.

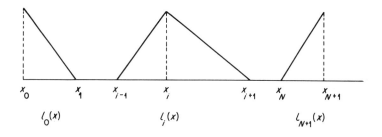

Figure 12

Example $m = 2$

The elements of $H^{(2)}(\Delta)$ are *piecewise cubic Hermite polynomials*. Writing $f_i^0 = f_i$, $\phi_{i0}(x) = v_i(x)$, $\phi_{i1}(x) = s_i(x)$, $i = 0, 1, \ldots, N+1$,

$$g(x) = \sum_{i=0}^{N+1} f_i v_i(x) + \sum_{i=0}^{N+1} f_i^1 s_i(x) \tag{8.5}$$

where the 'value functions' $v_i(x)$ are defined by

$$\left.\begin{array}{r} v_i(x_j) = \delta_{ij} \\ Dv_i(x_j) = 0 \end{array}\right\} \quad i, j = 0, 1, \ldots, N+1,$$

and the 'slope functions' $s_i(x)$ by

$$\left.\begin{array}{r} s_i(x_j) = 0 \\ Ds_i(x_j) = \delta_{ij} \end{array}\right\} \quad i, j = 0, 1, \ldots, N+1.$$

Letting $\alpha_i = (x - x_i)/h_{i-1}, \beta_i = (x - x_i)/h_i$, we have explicitly (see also Figures 13 and 14),

$$v_0(x) = \begin{cases} 1 + 2\beta_0^3 - 3\beta_0^2 & x_0 \leqslant x \leqslant x_1 \\ 0 & x_1 \leqslant x \leqslant x_{N+1} \end{cases}$$

$$v_i(x) = \begin{cases} 0 & x_0 \leqslant x \leqslant x_{i-1} \\ 1 - 3\alpha_i^2 - 2\alpha_i^3 & x_{i-1} \leqslant x \leqslant x_i \\ 1 - 3\beta_i^2 + 2\beta_i^3 & x_i \leqslant x \leqslant x_{i+1} \\ 0 & x_{i+1} \leqslant x \leqslant x_{N+1} \end{cases}$$

$$v_{N+1}(x) = \begin{cases} 0 & x_0 \leqslant x \leqslant x_N \\ 1 - 3\alpha_{N+1}^2 - 2\alpha_{N+1}^3 & x_N \leqslant x \leqslant x_{N+1} \end{cases}$$

$$s_0(x) = \begin{cases} h_0 \beta_0 (1 - \beta_0)^2 & x_0 \leqslant x \leqslant x_1 \\ 0 & x_1 \leqslant x \leqslant x_{N+1} \end{cases}$$

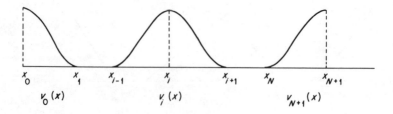

Figure 13

$$s_i(x) = \begin{cases} 0 & x_0 \leqslant x \leqslant x_{i-1} \\ h_{i-1}\alpha_i(1+\alpha_i)^2 & x_{i-1} \leqslant x \leqslant x_i \\ h_i\beta_i(1-\beta_i)^2 & x_i \leqslant x \leqslant x_{i+1} \\ 0 & x_{i+1} \leqslant x \leqslant x_{N+1} \end{cases}$$

$$s_{N+1}(x) = \begin{cases} 0 & x_0 \leqslant x \leqslant x_N \\ h_N\alpha_{N+1}(1+\alpha_{N+1})^2 & x_N \leqslant x \leqslant x_{N+1}. \end{cases}$$

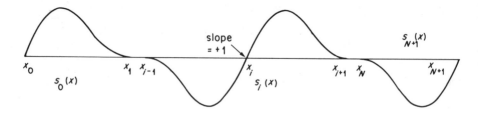

Figure 14

If the partition of the interval $[a, b]$ is uniform so that $h_i = h, i = 0, 1, \ldots, N$, then we can express all these basis functions in terms of two 'standard' basis functions $v(x)$ and $s(x)$. If we define

$$v(x) = \begin{cases} 0 & x \leqslant -1 \\ (x+1)^2(1-2x) & -1 \leqslant x \leqslant 0 \\ (1-x)^2(1+2x) & 0 \leqslant x \leqslant 1 \\ 0 & 1 \leqslant x \end{cases}$$

and

$$s(x) = \begin{cases} 0 & x \leqslant -1 \\ x(x+1)^2 & -1 \leqslant x \leqslant 0 \\ x(1-x)^2 & 0 \leqslant x \leqslant 1 \\ 0 & 1 \leqslant x \end{cases}$$

then it may be shown that

$$
\left.
\begin{aligned}
v_i(x) &= v\left(\frac{x - x_0}{h} - i\right) \\[2mm]
s_i(x) &= hs\left(\frac{x - x_0}{h} - i\right)
\end{aligned}
\right\} \quad i = 0, 1, \ldots, N + 1.
$$

The verification of this is left as an exercise.

Higher dimensional analogues of these functions exist, through the use of tensor products. In two dimensions, the case $m = 1$ gives rise to *piecewise bilinear functions*. Consider the region $[x_0, x_{N+1}] \times [y_0, y_{M+1}]$. Then if $g(x, y) \in H^{(1)}(\Delta)$, where Δ is now a two-dimensional rectangular grid of MN points in this region, we can write

$$
g(x, y) = \sum_{i=0}^{N+1} \sum_{j=0}^{M+1} f_{ij} \phi_{ij}(x, y) \tag{8.6}
$$

where $\phi_{ij}(x, y) = l_i(x) l_j(y)$, $i = 0, 1, \ldots, N + 1$, $j = 0, 1, \ldots, M + 1$, with $l_i(x)$ and $l_j(y)$ defined precisely as before with the obvious modification that $l_j(y)$ has $h_0, h_1, \ldots, h_N$ replaced by $k_0, k_1, \ldots, k_M$ where $k_i = y_{i+1} - y_i$, $i = 0, 1, \ldots, M$. Thus a typical basis function $\phi_{ij}(x, y)$ has support on the region $[x_{i-1}, x_{i+1}] \times [y_{j-1}, y_{j+1}]$, with values as shown in Figure 15.

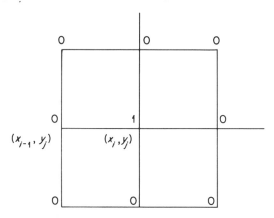

Figure 15

In two dimensions, $H^{(2)}(\Delta)$ is the space of *piecewise bicubic Hermite polynomials*. With $v_i(x)$, $v_j(y)$, $s_i(x)$, $s_j(y)$ defined as before (with obvious modifications to the functions of y) we can write $g(x, y) \in H^{(2)}(\Delta)$ as

$$
\begin{aligned}
g(x, y) = \sum_{i=0}^{N+1} \sum_{j=0}^{M+1} \{ &a_{ij} v_i(x) v_j(y) + b_{ij} s_i(x) v_j(y) \\
&+ c_{ij} v_i(x) s_j(y) + d_{ij} s_i(x) s_j(y) \},
\end{aligned}
$$

which satisfies the conditions (in an obvious notation)

$$g(x_i, y_j) = a_{ij}$$
$$D_x g(x_i, y_j) = b_{ij}$$
$$D_y g(x_i, y_j) = c_{ij}$$
$$D_x D_y g(x_i, y_j) = d_{ij},$$

for $i = 0, 1, \ldots, N + 1, j = 0, 1, \ldots, M + 1$. A typical basis function has support on the region $[x_{i-1}, x_{i+1}] \times [y_{j-1}, y_{j+1}]$.

Exercises

1. Complete the details of the proof of Theorem 8.1.

2. Let $m = 1$, let $f_i^0 = f(x_i)$, $i = 0, 1, \ldots, N + 1$, and assume that the points x_i are equispaced. Show that integrating the piecewise linear Hermite interpolant satisfying (8.1) gives the trapezoidal rule for approximating $\int_a^b f(x)\, dx$. If $m = 2$, and $f_i^1 = Df(x_i)$, $i = 0, 1, \ldots, N + 1$, what integration formula is the result of integrating the piecewise cubic Hermite interpolant?

3. Verify that the basis functions can be expressed in terms of the 'standard' basis functions introduced in this section for $m = 1$ and 2.

4. Let $f_{ij} = f(x_i, y_j)$, $i = 0, 1, \ldots, N + 1, j = 0, 1, \ldots, M + 1$. For fixed y, let $p_x f$ denote the piecewise linear Hermite interpolant to $f_i = f(x_i, y)$, $i = 0, 1, \ldots, N + 1$ (we may regard p_x as an operator) and for fixed x, let $p_y f$ denote the piecewise linear Hermite interpolant to $f_j = f(x, y_j)$, $j = 0. 1, \ldots, M + 1$. Prove that the piecewise *bilinear* interpolant $p_{xy} f$ defined by equation (8.6) satisfies

$$p_{xy} f = p_x p_y f = p_y p_x f.$$

5. Prove that $p_x f$ defined in the previous exercise satisfies

$$\| p_x f \|_\infty \leqslant \max_{0 \leqslant i \leqslant N+1} |f_i|.$$

6. Prove the analogue of the result of exercise 4 for the *cubic* Hermite interpolant.

7. Let $x_0 = 0$, $x_{N+1} = 1$, $h = x_{i+1} - x_i$, $i = 0, 1, \ldots, N$. Prove that

$$|s_i(x)| + |s_{i+1}(x)| \leqslant \tfrac{1}{4} h$$

for any $x \in [0, 1]$, where $s_i(x) \in H^{(2)}(\Delta)$ is the slope function.

8.3 Error bounds for $H^{(2)}(\Delta)$ interpolants

We turn now to the problem of providing *a priori* error bounds for the interpolation procedure introduced in the previous section for the one-dimensional cubic Hermite case, when f_i and f_i^1 correspond to function value and derivative of $f(x)$ at $x = x_i$. Although the analysis is carried out for $H^{(2)}(\Delta)$, precisely analogous results hold for other values of m; in particular, some of these for the (simpler) case $m = 1$ are covered in the exercises. The error bounds, and their higher dimensional analogues, are most useful in an analysis of finite element methods for the approximate solution of differential equations, in which the exact solution may be regarded as being represented by a piecewise polynomial. It is beyond the scope of this text to deal with the finite element method in any detail; however, some motivation for subsequent analysis is provided by an illustration of its application (see Mitchell and Wait, 1977, for a good introduction to the method).

Consider the differential equation

$$Au = f \qquad \mathbf{x} \in R$$

with $u = 0$ on ∂R, the boundary of R, where A is a second order linear differential operator. Then if A is positive definite and self-adjoint, u is a solution if and only if $u = 0$ on ∂R and in addition

$$\langle Au, v \rangle = \langle f, v \rangle$$

for all functions v which also satisfy the boundary condition, and for which

$$\langle v, v \rangle \equiv \int_R v^2 \, d\mathbf{x} = \| v \|_2^2$$

is defined. For a wide class of such problems, use of the boundary condition and integration by parts enables the equation for u to be written as

$$l(u, v) = \langle f, v \rangle$$

where l is a particular bilinear form. For example, in one dimension if $A = -D^2$, then $l(u, v) \equiv \langle Du, Dv \rangle$. The finite element method proceeds by representing $u(\mathbf{x})$ by an element of a piecewise polynomial subspace, and satisfying this relation for all members v of the subspace. If $u(\mathbf{x})$ is approximated by

$$U(\mathbf{x}) = \sum_{i=1}^{n} a_i \phi_i(\mathbf{x})$$

where $\phi_1(\mathbf{x}), \phi_2(\mathbf{x}), \ldots, \phi_n(\mathbf{x})$ is a basis for the subspace, we have to satisfy the system of linear equations

$$\sum_{i=1}^{n} a_i l(\phi_i, \phi_j) = \langle f, \phi_j \rangle \qquad j = 1, 2, \ldots, n.$$

The matrix of the left hand side of this linear system of equations for $\mathbf{a}$ is symmetric if A is self-adjoint and will contain inner products of the form $\langle D\phi_i, D\phi_j \rangle$ and $\langle \phi_i, \phi_j \rangle$. A consequence of this is that compact support of the basis functions is likely to lead to a matrix which has the nonzero elements mainly in a band down the main diagonal, and this is a desirable structure from a numerical point of view.

Now let the subspace of approximating functions be $H^{(2)}(\Delta)$ and let $p_H f$ denote the unique element of $H^{(2)}(\Delta)$ which is defined by interpolation to the numbers $f_i \equiv f_i^0, f_i^1, i = 0, 1, \ldots, N+1$, as in Theorem 8.1. If there is an underlying function $f(x)$ defined and continuously differentiable in $[x_0, x_{N+1}]$, then we will assume that $f_i = f(x_i)$, $f_i^1 = Df(x_i)$, $i = 0, 1, \ldots, N+1$, and can regard p_H as an operator. Then it is often possible to obtain a bound for $\|u(x) - U(x)\|_2$ of the form

$$\|u(x) - U(x)\|_2 \leqslant K \|D(u - p_H u)\|_2 \|D(w - p_H w)\|_2$$

where K is a constant, and $w \in C^1[x_0, x_{N+1}]$ depends on the differential equation (see, for example, Schultz, 1973). The provision of an L_2 bound for the derivative of the error in the Hermite interpolant is therefore an important requirement in such an analysis.

The derivation of appropriate L_2 error bounds will not be presented in full detail here. Such bounds are, however, a consequence of the interpolant possessing a certain best approximation property, and this will be examined. We will then derive an L_∞ error bound which, in addition to being of interest for its own sake, serves also to illustrate the order of accuracy of the approximation.

For given numbers f_i, f_i^1, $i = 0, 1, \ldots, N+1$, let W_f denote the set

$$W_f = \{w(x): w(x) \in C^1[a, b], w(x) \in C^2(x_i, x_{i+1}), i = 0, 1, \ldots, N,$$
$$w(x_i) = f_i, Dw(x_i) = f_i^1, i = 0, 1, \ldots, N+1\},$$

and let W_0 be the set W_f with $f_i = f_i^1 = 0, i = 0, 1, \ldots, N+1$.

Theorem 8.2 $p_H f$ is the unique solution to the problem: find $w \in W_f$ to minimize $\|D^2 w\|_2^2$.

Proof Clearly $p_H f \in W$. Let $w \in W$ be arbitrary. Then

$$\|D^2 w - D^2 p_H f\|_2^2 = \|D^2 w\|_2^2 - \|D^2 p_H f\|_2^2 + 2\langle D^2 p_H f - D^2 w, D^2 p_H f \rangle. \tag{8.7}$$

Let $w_0 = p_H f - w$. Then $w_0 \in W_0$, and so

$$\langle D^2 w_0, D^2 p_H f \rangle = \int_a^b D^2 w_0 \, D^2 p_H f \, dx$$

$$= \sum_{i=0}^N \int_{x_i}^{x_{i+1}} D^2 w_0 D^2 p_H f \, dx$$

$$= \sum_{i=0}^N [Dw_0 D^2 p_H f]_{x_i}^{x_{i+1}} - \sum_{i=0}^N \int_{x_i}^{x_{i+1}} Dw_0 D^3 p_H f \, dx$$

$$= -\sum_{i=0}^N [w_0 D^3 p_H f]_{x_i}^{x_{i+1}} + \sum_{i=0}^N \int_{x_i}^{x_{i+1}} w_0 D^4 p_H f \, dx$$

$$= 0,$$

where we have integrated by parts twice, and used the interpolation conditions. Thus (8.7) gives

$$\|D^2 w\|_2^2 - \|D^2 p_H f\|_2^2 = \|D^2(w - p_H f)\|_2^2 \tag{8.8}$$
$$\geqslant 0;$$

it remains to show that the minimizer is unique. Assume that there exists $w \in W_f$ such that $\|D^2 w\|_2^2 = \|D^2 p_H f\|_2^2$. Then from (8.8),

$$\|D^2(w - p_H f)\|_2^2 = 0,$$

and so $D^2(w - p_H f) = 0$, $x \in [x_i, x_{i+1}]$, $i = 0, 1, \ldots, N$. Thus $w_0 = w - p_H f$ is a linear function of x in these subintervals, and since $w_0 \in W_0$ it follows that we must have $w = p_H f$. $\square$

Theorem 8.2 may be used as a basis for the computation of the following L_2 error bounds (see, for example, Schultz, 1973, for full details). Provided that $f(x)$ is sufficiently differentiable, we have

$$\|D^k(f - p_H f)\|_2 \leqslant \pi^{k-4} h^{4-k} \|D^4 f\|_2 \qquad k = 0, 1, 2 \tag{8.9}$$

where $h = \max_i h_i$. Two-dimensional analogues of this result are also available, for example

$$\|f - p_H f\|_2 \leqslant \pi^{-4}(h_1^4 \|D_x^4 f\|_2 + h_1^2 h_2^2 \|D_x^2 D_y^2 f\|_2 + h_2^4 \|D_y^4 f\|_2)$$

where h_1 and h_2 are respectively the maximum intervals in the x and y directions, and $p_H f$ is redefined appropriately.

We now derive the following L_∞ error bound.

Theorem 8.3 If $f(x) \in C^4[a, b]$, then

$$\|f - p_H f\|_\infty \leqslant \tfrac{1}{384} h^4 \|D^4 f\|_\infty. \tag{8.10}$$

Proof In the interval $[x_i, x_{i+1}]$, we are fitting a cubic Hermite interpolating polynomial. Thus (exercise 29 of Chapter 1)

$$\max_{x \in [x_i, x_{i+1}]} |f - p_H f| \leqslant \frac{1}{4!} \max_{x \in [x_i, x_{i+1}]} |D^4 f| \max_{x \in [x_i, x_{i+1}]} |(x - x_i)^2 (x - x_{i+1})^2|$$

$$\leqslant \frac{1}{4!} \frac{h^4}{2^4} \max_{x \in [x_i, x_{i+1}]} |D^4 f|.$$

Since this is true for all i, $i = 0, 1, \ldots, N$, the result follows immediately. $\square$

Corollary

$$\|f - p_H f\|_2 \leqslant \frac{\sqrt{(b-a)}}{384} h^4 \|D^4 f\|_\infty.$$

Similar L_∞ error bounds may also be obtained for derivatives. It may be shown that

$$\| D^k(f - p_H f)\|_\infty \leqslant C_k h^{4-k} \| D^4 f\|_\infty \qquad k = 1, 2, 3$$

where C_k is a constant; for details see Birkhoff and Priver (1967). Similar results may also be obtained in two dimensions: corresponding to equation (8.10), it may be shown that

$$\| f - p_H f\|_\infty \leqslant \frac{1}{384} h_1{}^4 \| D_x{}^4 f\|_\infty + \frac{1}{16} h_1{}^2 h_2{}^2 \| D_x{}^2 D_y{}^2 f\|_\infty + \frac{1}{384} h_2{}^4 \| D_y{}^4 f\|_\infty$$

$$(8.11)$$

where h_1 and h_2 are as before (Schultz, 1973). The bounds (8.9) (with $k = 0$), (8.10), and (8.11) are usually said to imply that the cubic Hermite interpolant is *fourth order accurate*.

Exercises

8. Let $X = [0, 1]$ and consider the differential equation

$$-D[p(x)Du(x)] + q(x)u(x) = f(x) \qquad 0 \leqslant x \leqslant 1$$
$$u(0) = u(1) = 0.$$

Obtain the form of $l(u, v)$ in this case. Obtain also the form of the matrices of the system of equations defining the finite element solution for the spaces $H^{(1)}(\Delta)$ and $H^{(2)}(\Delta)$ respectively.

9. Let V_0 be the set of elements $v(x)$ whose L_2 norm exists on $[0, 1]$ and which satisfy $v(0) = v(1) = 0$, and assume that the following are equivalent
    (i) $u(x)$ solves the differential equation in exercise 8.
    (ii) $u(x)$ satisfies $u(0) = u(1) = 0$ and also $l(u, v) = \langle f, v \rangle$ for all $v \in V_0$.
Prove that $u(x)$ is the unique solution to the problem:

$$\text{find } w \in V_0 \text{ to minimize } F(w) \equiv l(w, w) - 2\langle w, f \rangle.$$

Show the connection between the solution to this problem and the finite element procedure.

10. Given numbers $f_i \equiv f_i^0$, $i = 0, 1, \ldots, N + 1$, let V_f denote the set

$$V_f = \{v(x): v(x) \in C[a, b], v(x) \in C^1(x_i, x_{i+1}), i = 0, 1, \ldots, N,$$
$$v(x_i) = f_i, i = 0, 1, \ldots, N + 1\},$$

and let $p_H f \in H^{(1)}(\Delta)$ be the interpolant defined by Theorem 8.1. Prove that $p_H f$ is the unique solution to the problem:

$$\text{find } v \in V_f \text{ to minimize } \| Dv\|_2{}^2.$$

11. Let $f(x) \in C^2[a, b]$ and let $p_H f$ be defined as in the previous exercise. Prove that

$$\| f - p_H f\|_\infty \leqslant \frac{1}{8} h^2 \| D^2 f\|_\infty$$

where $h = \max_i h_i$. (This shows that the linear Hermite interpolant is second order accurate.)

8.4 The spline space $S^{(n)}(\Delta)$

We assume that X and Δ are as defined in Section 8.2.

Definition 8.2   The spline space $S^{(n)}(\Delta)$ is given by

$$S^{(n)}(\Delta) = \{g(x) \in C^{n-1}[X]; \; g(x) \text{ is a polynomial of degree } n \text{ on each interval } [x_i, x_{i+1}], \; i = 0, 1, \ldots, N, \text{ defined by } \Delta\}.$$

This class of approximating functions thus consists of piecewise polynomials of degree n which have $(n-1)$ continuous derivatives at the knots x_i, $i = 1, 2, \ldots, N$. Thus $S^{(n)}(\Delta)$ differs from $H^{(m)}(\Delta)$ essentially in that additional continuity is required at the knots. It is clear that if $n = 2m - 1$, $S^{(n)}(\Delta) \subset H^{(m)}(\Delta)$ with $S^{(1)}(\Delta) \equiv H^{(1)}(\Delta)$. Originally, a 'spline' referred to a physical device used by draughtsmen for drawing smooth curves, consisting of a long flexible rod which could be bent to pass through any required points.

The dimension of $S^{(n)}(\Delta)$ is easily seen to be $N + n + 1$, and any element $g(x)$ of $S^{(n)}(\Delta)$ can be uniquely defined by an interpolation procedure analogous to that for elements of $H^{(m)}(\Delta)$. The following result is valid for splines of *odd degree*.

Theorem 8.4   Let $m \leqslant N + 2$. Then for any prescribed set of values f_i, $i = 0, 1, \ldots, N + 1$, and $f_0{}^k, f_{N+1}{}^k, k = 1, 2, \ldots, m - 1$, there is a unique element $g(x) \in S^{(n)}(\Delta)$, where $n = 2m - 1$, such that

$$\begin{aligned} g(x_i) &= f_i & i &= 0, 1, \ldots, N + 1 \\ D^k g(x_i) &= f_i^k & i &= 0, N + 1; \; k = 1, 2, \ldots, m - 1. \end{aligned} \tag{8.12}$$

Proof   Given any set of $N + n + 1$ basis functions for $S^{(n)}(\Delta)$, the conditions (8.12) give a linear system of $N + n + 1$ equations for the $N + n + 1$ coefficients defining $g(x)$. The required result will follow if the system is nonsingular, or alternatively if $g(x) = 0$ is the only element of $S^{(n)}(\Delta)$ satisfying (8.12) with $f_i = 0, i = 0, 1, \ldots, N + 1$, $f_i^k = 0, i = 0, N + 1; k = 1, 2, \ldots, m - 1$.

If $g(x)$ satisfies this system of equations, it follows from exercise 14 that

$$\int_a^b [D^m g(x)]^2 \, dx = 0$$

so that $D^m g(x)$ vanishes identically on $[a, b]$. Thus $g(x)$ is a polynomial of degree $(m - 1)$ or less, and since it has $N + 2 > m - 1$ zeros it must be identically zero. $\square$

For odd degree splines, basis functions analogous to those introduced for $H^{(m)}(\Delta)$ it may be defined in a similar manner. These are called *cardinal splines*, and for $n = 3$ $(m = 2)$ they are defined as follows

$$\left.\begin{array}{l} C_i(x_j) = \delta_{ij} \\ DC_i(x_0) = DC_i(x_{N+1}) = 0 \end{array}\right\} \quad i, j = 0, 1, \ldots, N+1$$

$$C_{N+2}(x_j) = C_{N+3}(x_j) = 0 \qquad j = 0, 1, \ldots, N+1$$

$$DC_{N+2}(x_0) = DC_{N+3}(x_{N+1}) = 1$$

$$DC_{N+2}(x_{N+1}) = DC_{N+3}(x_0) = 0.$$

Then $C_i(x) \in S^{(3)}(\Delta)$, $i = 0, 1, \ldots, N+3$, by Theorem 8.4, and for any $g(x)$ defined by this theorem we can write

$$g(x) = \sum_{i=0}^{N+1} f_i C_i(x) + f_0{}^1 C_{N+2}(x) + f_{N+1}{}^1 C_{N+3}(x). \tag{8.13}$$

Unlike the basis functions introduced for $H^{(m)}(\Delta)$, the cardinal splines do not have compact support (the graph of a typical $C_i(x)$ is shown in Figure 16). This means that if the cardinal splines are used as a basis for *any* cubic spline, so that $g(x) \in S^{(3)}(\Delta)$ is represented as

$$g(x) = \sum_{i=0}^{N+1} a_i C_i(x) + a_{N+2} C_{N+2}(x) + a_{N+3} C_{N+3}(x) \tag{8.14}$$

then there are two major disadvantages. Firstly, the determination of the coefficients a_i, $i = 0, 1, \ldots, N+3$ by any means other than interpolation at the knots will result, in general, in a system of linear equations to be solved which has a full (dense) matrix; and secondly, evaluation of $g(x)$ at an arbitrary point involves all terms of the sum. This contrasts with the analogous situation in the Hermite case, when the restricted support results in a matrix which consists of zeros except in a band down the main diagonal: the coefficients may be more efficiently obtained and subsequent manipulation of the expression is easier.

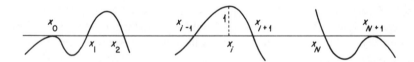

Figure 16

Thus, from a numerical point of view, the ideal basis functions should have compact support, with the interval of the support being reasonably small. Such a set of functions is provided by *B-splines*, which are actually the basis functions of minimal support for a given degree. To define these splines (for general n), it is convenient to extend Δ by adding $2n$ additional points outside X such that

$$x_{-n} < x_{-n+1} < \cdots < x_0 < x_1 < \cdots < x_N < x_{N+1} < \cdots < x_{N+n+1}.$$

Definition 8.3 The nth divided difference of the function $f(x; y)$ on the $(n+1)$ points $y_0, y_1, \ldots, y_n$ is given by the recursive formula

$$f(x; y_0, y_1, y_2, \ldots, y_n) = \frac{f(x; y_1, y_2, \ldots, y_n) - f(x; y_0, y_1, \ldots, y_{n-1})}{y_n - y_0}.$$

Definition 8.4 For $i = 1, 2, \ldots, N+n+1$, the ith B-spline of degree n is given by

$$M_{ni}(x) = M(x; x_{i-n-1}, x_{i-n}, \ldots, x_i), \tag{8.15}$$

with

$$M(x; y) = (y - x)_+^n$$

where

$$x_+^n = \begin{cases} x^n & x \geqslant 0 \\ 0 & x < 0 \end{cases}.$$

In terms of the function

$$w_{ni}(x) = (x - x_{i-n-1})(x - x_{i-n}) \ldots (x - x_i)$$

an explicit expression for $M_{ni}(x)$ is given by (see exercise 17)

$$M_{ni}(x) = \sum_{j=i-n-1}^{i} \frac{(x_j - x)_+^n}{Dw_{ni}(x_j)}. \tag{8.16}$$

The B-splines defined on a uniform partition (in which $x_{i+1} - x_i = h$, $i = -n, \ldots, N+n+1$) were first proposed by Schoenberg in 1946, and given for the general case by Curry and Schoenberg (1947). The property of compact support may be deduced directly from (8.15) and (8.16): for $x \geqslant x_i$, all terms of the sum in (8.16) are zero; for $x \leqslant x_{i-n-1}$, $M(x; y)$ is just a polynomial of degree n, whose $(n+1)$st divided difference is therefore zero. Thus $M_{ni}(x)$ has support confined to the interval $[x_{i-n-1}, x_i]$.

Example $n = 3$

Assuming that the partition is uniform, we will construct the cubic B-splines from their definition (8.15). We have

$$M_{3i}(x) = M(x; x_{i-4}, \ldots, x_i)$$
$$= \frac{1}{h^4} \frac{\delta^4 M(x; x_{i-2})}{4!}$$

where δ is the central difference operator (exercise 18)

$$= \frac{1}{h^4} \frac{\delta^4 [(x_{i-2} - x)_+^3]}{4!}$$

$$= \frac{1}{4!h^4}[(x_i - x)_+^3 - 4(x_{i-1} - x)_+^3 + 6(x_{i-2} - x)_+^3$$
$$- 4(x_{i-3} - x)_+^3 + (x_{i-4} - x)_+^3].$$

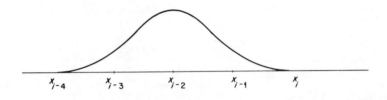

$$x_{i-4} \qquad x_{i-3} \qquad x_{i-2} \qquad x_{i-1} \qquad x_i$$

Figure 17

The graph of $M_{3i}(x)$ is shown in Figure 17.

Curry and Schoenberg (1966) show that $M_{ni}(x)$ is strictly positive on (x_{i-n-1}, x_i), and also that any $g(x) \in S^{(n)}(\Delta)$ has a unique representation

$$g(x) = \sum_{i=1}^{N+n+1} a_i M_{ni}(x). \tag{8.17}$$

A neat proof of this latter result is given by de Boor and Fix (1973). Defining the linear functionals λ_i by

$$\lambda_i(g(x)) = \frac{(x_i - x_{i-n-1})}{n!} \sum_{r=0}^{n} (-1)^{n-r} (D^{n-r}\psi_i)(\xi_i)(D^r g)(\xi_i)$$
$$i = 1, 2, \ldots, N+n+1,$$

where $g(x) \in S^{(n)}(\Delta)$, $\psi_i(x) = (x_{i-n} - x) \ldots (x_{i-1} - x)$, and ξ_i satisfies $x_{i-n-1} < \xi_i < x_i$ (with ξ_i replaced by $\xi_i +$ if necessary), they show that

$$\lambda_j(M_{ni}(x)) = \delta_{ij} \qquad i, j = 1, 2, \ldots, N+n+1.$$

It follows that the set of B-splines is linearly independent on $[x_0, x_{N+1}]$, and since $M_{ni}(x) \in S^n(\Delta)$, $i = 1, 2, \ldots, N+n+1$, they form a basis for $S^{(n)}(\Delta)$. A further consequence is that in the representation (8.17),

$$a_i = \lambda_i(g(x)) \qquad i = 1, 2, \ldots, N+n+1.$$

The B-spline basis is now accepted as being the appropriate basis for numerical work. That being so, it is clearly important that efficient, numerically accurate methods for the computation of B-spline values be available. It is known that the conventional method of evaluating B-splines using the formula (8.16) can lead to a numerically unstable process, and an alternative stable procedure has been given by Cox (1972), based on the following result.

Theorem 8.5 The recurrence relation

$$M_{ni}(x) = \frac{(x - x_{i-n-1}) M_{n-1, i-1}(x) + (x_i - x) M_{n-1, i}(x)}{x_i - x_{i-n-1}} \tag{8.18}$$

is valid for all values of x.

Proof The right hand side of (8.18) is, using (8.16),

$$\frac{x - x_{i-n-1}}{x_i - x_{i-n-1}} \sum_{j=i-n-1}^{i-1} \frac{(x_j - x)_+^{n-1}}{Dw_{n-1,i-1}(x_j)} + \frac{(x_i - x)}{x_i - x_{i-n-1}} \sum_{j=i-n}^{i} \frac{(x_j - x)_+^{n-1}}{Dw_{n-1,i}(x_j)}$$

$$= \frac{(x - x_{i-n-1})(x_{i-n-1} - x)_+^{n-1}}{(x_i - x_{i-n-1})Dw_{n-1,i-1}(x_{i-n-1})}$$

$$+ \sum_{j=i-n}^{i-1} \frac{(x_j - x)_+^{n-1}}{x_i - x_{i-n-1}} \left(\frac{x - x_{i-n-1}}{Dw_{n-1,i-1}(x_j)} + \frac{x_i - x}{Dw_{n-1,i}(x_j)} \right)$$

$$+ \frac{(x_i - x)(x_i - x)_+^{n-1}}{(x_i - x_{i-n-1})Dw_{n-1,i}(x_i)}. \tag{8.19}$$

The following identities

$$Dw_{n,i}(x_j) = (x_j - x_i)Dw_{n-1,i-1}(x_j) \qquad j \neq i$$

$$Dw_{n,i}(x_j) = (x_j - x_{i-n-1})Dw_{n-1,i}(x_j) \qquad j \neq i-n-1$$

which follow from the definition of $w_{ni}(x)$, may be used to replace the first and last terms on the right hand side of (8.19) by, respectively,

$$\frac{(x_{i-n-1} - x)_+^n}{Dw_{ni}(x_{i-n-1})} \quad \text{and} \quad \frac{(x_i - x)_+^n}{Dw_{ni}(x_i)}. \tag{8.20}$$

In addition, the jth term of the summation in (8.19) may be written as

$$\frac{(x_j - x)_+^{n-1}\{(x_j - x_i)(x - x_{i-n-1}) + (x_j - x_{i-n-1})(x_i - x)\}}{(x_i - x_{i-n-1})Dw_{ni}(x_j)} = \frac{(x_j - x)_+^n}{Dw_{ni}(x_j)}.$$

Summing this term over all j from $i - n$ to $i - 1$ and adding the terms in (8.20) gives

$$\sum_{j=i-n-1}^{i} \frac{(x_j - x)_+^n}{Dw_{ni}(x_j)}$$

which is just $M_{ni}(x)$ by (8.16). ☐

If the B-splines of degree $n = 0$ are calculated for $j = i - n, i - n + 1, \ldots, i$, then $M_{ni}(x)$ may be obtained using the recurrence relation (8.18). For example, for the case $n = 3$ we will obtain the following triangular array

$$
\begin{array}{cccc}
M_{0,i-3} & & & \\
 & M_{1,i-2} & & \\
M_{0,i-2} & & M_{2,i-1} & \\
 & M_{1,i-1} & & M_{3,i}. \\
M_{0,i-1} & & M_{2,i} & \\
 & M_{1,i} & & \\
M_{0,i} & & &
\end{array}
$$

Advantage may be taken of zero elements in the array; for example since the support of the degree 0 B-spline $M_{0,i}$ is confined to $[x_{i-1}, x_i]$ (explicitly $M_{0,i}(x) = 1/(x_i - x_{i-1})$, $x_{i-1} \leqslant x \leqslant x_i$) then if x lies outside this interval, the values of $M_{0,i}(x)$ will be identically zero. For example if x lies in the interval $[x_{i-2}, x_{i-1}]$, then the above triangular array takes the form

$$
\begin{array}{ccccc}
0 & & & & \\
 & 0 & & & \\
0 & & & & M_{2,i-1} \\
 & & M_{1,i-1} & & & M_{3,i}. \\
 & M_{0,i-1} & & M_{2,i} & \\
 & & M_{1,i} & & \\
 & 0 & & &
\end{array}
$$

The nonzero values, therefore, will form a rhomboidal, rather than a triangular array. Cox (1972) proposes the following algorithm, which takes full advantage of zero entries. We assume that a value of x is given satisfying $x_{l-1} \leqslant x \leqslant x_l$.

(1)   Set $k = i - l$,
$$M_{r,l-1}(x) = 0 \qquad r = 0, 1, \ldots, n - k - 1$$
$$M_{r,l+r+1}(x) = 0 \qquad r = 0, 1, \ldots, k - 1.$$
(2)   Compute $M_{r,l+j}(x)$ from (8.18) for
$$r = j, j+1, \ldots, j + n - k$$
$$j = 0, 1, \ldots, k,$$
except for the case $r = j = 0$ when $M_{0l}(x)$ is set to $1/(x_l - x_{l-1})$.

The total number of storage locations required is at most $n + 1$, since as soon as $M_{r,l+j}$ has been computed, it may overwrite $M_{r-1,l+j-1}$. Since all the individual quantities occurring on the right hand side of (8.18) are non-negative there will be no cancellation in its evaluation, and thus we can expect the above algorithm to be numerically stable. Detailed error analyses are given by Cox (1972) to show that this is indeed the case. He shows that the proposed method possesses advantages in both speed and accuracy over the conventional method which uses (8.16) directly. The evaluation of B-splines, and of a spline from its B-spline representation, has also been considered by de Boor (1972) and Cox (1978b).

The condition of the B-spline basis has been investigated by de Boor (1972) and Lyche (1978). Let the B-splines be normalized so that we have

$$B_{ni}(x) = (x_i - x_{i-n-1})M_{ni}(x) \qquad i = 1, 2, \ldots, N + n + 1,$$

and let

$$g(x, \mathbf{a}) = \sum_{i=1}^{N+n+1} a_i B_{ni}(x).$$

Defining

$$K_{n,\Delta} = \sup_{\|\mathbf{a}\|_\infty = 1} \|g(x, \mathbf{a})\|_\infty \Big/ \inf_{\|\mathbf{a}\|_\infty = 1} \|g(x, \mathbf{a})\|_\infty$$

for a given set of knots defined by Δ, and

$$K_n = \sup_{N+n+1 \geqslant 1} \sup_{\Delta} K_{n,\Delta}$$

where the supremum is taken over all possible Δ, then Lyche (1978) shows that

$$\frac{n}{n+1} 2^{n-\frac{1}{2}} \leqslant K_n \leqslant 2(n+1)9^n \qquad n \geqslant 1.$$

Thus the condition number measured by K_n increases exponentially with the degree n.

It is often necessary to calculate derivatives of B-splines (for example when an approximate solution to a differential equation is sought in the form (8.17)), and stable numerical methods for this also exist; see Cox (1978a) and Butterfield (1976). The calculation of integrals of B-splines is considered by Gaffney (1976) and de Boor, Lyche, and Schumaker (1976).

Higher dimensional analogues of the B-splines defined here can be constructed through the use of tensor products. In two dimensions, consider the region $[x_0, x_{N+1}] \times [y_0, y_{M+1}]$ and let Δ be a rectangular grid of MN points in this region. Then if $N_{n,j}(y)$, $j = 1, 2, \ldots, M+n+1$, are B-splines defined on the knots $\{y_j\}$, a general degree n bivariate spline may be represented as

$$g(x, y) = \sum_{i=1}^{N+n+1} \sum_{j=1}^{M+n+1} a_{ij} M_{n,i}(x) N_{n,j}(y);$$

the basis spline $M_{n,i}(x)N_{n,j}(y)$ has support confined to the region $[x_{i-n-1}, x_i] \times [y_{j-n-1}, y_j]$.

Exercises

12. A function $g(x)$ is defined on the real line by

$$\begin{aligned}
g(x) &= 1 - 2x & x &\leqslant -3 \\
&= x^3 + 9x^2 + 25x + 28 & -3 &\leqslant x \leqslant -1 \\
&= -x^3 + 3x^2 + 19x + 26 & -1 &\leqslant x \leqslant 0 \\
&= -2x^3 + 3x^2 + 19x + 26 & 0 &\leqslant x \leqslant 3 \\
&= 5x^3 - 60x^2 + 208x - 163 & 3 &\leqslant x \leqslant 4 \\
&= 157 - 32x & 4 &\leqslant x.
\end{aligned}$$

Verify that $g(x)$ is a cubic spline function.

13. Prove that $g(x) \in S^{(n)}(\Delta)$ if and only if it can be expressed in the form

$$g(x) = \sum_{i=0}^{n} a_i x^i + \sum_{i=1}^{N} c_i (x - x_i)_+^n$$

where $c_i = [D^n g(x_i+) - D^n g(x_i-)]/n!$, $i = 1, 2, \ldots, N$. Express the cubic spline of exercise 12 in this form. (This is a convenient representation of an element of $S^{(n)}(\Delta)$ for *theoretical purposes*.)

14. Let $n = 2m - 1$ and let $f(x) \in C^{m-1}[a, b]$, $f(x) \in C^m(x_i, x_{i+1})$, $i = 0, 1, \ldots, N$ with

$$f(x_j) = 0 \qquad\qquad j = 0, 1, \ldots, N+1$$
$$D^k f(x_0) = D^k f(x_{N+1}) = 0 \qquad k = 1, 2, \ldots, m-1.$$

Prove that

$$\int_a^b D^m f(x) D^m g(x)\, dx = 0$$

for any $g(x) \in S^{(n)}(\Delta)$.

15. Determine the quadratic spline with knots (a) $x_1 = -1$, $x_2 = +1$ (b) $x_1 = -\frac{1}{2}$, $x_2 = \frac{1}{2}$ which interpolates the data

x	$-\alpha$	$-\frac{1}{2}$	0	$\frac{1}{2}$	α
f	0	0	ε	0	0

16. Derive the cardinal splines of degree 1 with knots $1, 2$ in the interval $[0, 3]$.

17. Use Definition 8.3 and recursion on n to prove that

$$f(x; x_0, x_1, \ldots, x_n) = \sum_{k=0}^n \frac{f(x; x_k)}{(x_k - x_0)\ldots(x_k - x_{k-1})(x_k - x_{k+1})\ldots(x_k - x_n)}.$$

Deduce equation (8.16).

18. The central difference operator δ is defined by

$$\delta f(x; x_i) = f(x; x_i + \tfrac{1}{2}h) - f(x; x_i - \tfrac{1}{2}h)$$

where the points x_i are equispaced, with interval h. In this case prove that

$$f(x; x_0, x_1, \ldots, x_n) = \frac{1}{h^n} \frac{\delta^n f(x; x_{n/2})}{n!}.$$

19. (a) Verify that

$$M_{3i}(x_{i-3}) = M_{3i}(x_{i-1}) = \frac{1}{24h}, \qquad\qquad M_{3i}(x_{i-2}) = \frac{1}{6h},$$

$$DM_{3i}(x_{i-3}) = -DM_{3i}(x_{i-1}) = \frac{1}{8h^2}, \qquad DM_{3i}(x_{i-2}) = 0$$

where $M_{3i}(x)$ is a cubic B-spline, defined on equispaced knots, with interval h.
(b) Derive the matrix of the system of linear equations obtained in approximating $f(x)$ by a cubic spline (in B-spline representation) by interpolating $f(x)$ at the N equispaced knots and at the end points, and interpolating derivatives at the end points, verifying that with the exception of 2 elements, it can be tridiagonal. What

essential difference will there be in the structure of the matrix if the knots are not equispaced?

20. For functions of 2 variables, a cubic spline interpolant analogous to that of equation (8.13) can be obtained. Write this out in full, and prove the result corresponding to exercise 4.

21. Let the B-splines be normalized so that

$$B_{n,i}(x) = (x_i - x_{i-n-1})M_{n,i}(x) \qquad i = 1, 2, \ldots, N+n+1.$$

Prove that $\sum_{i=1}^{N+n+1} B_{n,i}(x) = 1$ (de Boor, 1972).

8.5 Error bounds for $S^{(3)}(\Delta)$ interpolants

Splines of odd degree are particularly useful as approximating functions, and one reason for this is that they possess a best approximation property analogous to that of Theorem 8.2. This may be used to derive L_2 error bounds similar to those already described in the Hermite case, with similar applications to the finite element method as indicated for Hermite functions at the start of Section 8.3. In particular, cubic splines are commonly used: they are of low degree, yet because second derivatives are continuous, they give rise to physically smooth curves. We will only consider the cubic case, therefore, although analogous results (as in the Hermite case) carry over to higher values of n.

For given numbers f_i, $i = 0, 1, \ldots, N+1$, $f_0{}^1$, $f_{N+1}{}^1$, let $p_s f \in S^{(3)}(\Delta)$ be the interpolant defined by (8.12), and let U_f denote the set

$$U_f = \{u(x) : u(x) \in C^1[a, b], \ u(x) \in C^2(x_i, x_{i+1}), \ i = 0, 1, \ldots, N,$$
$$u(x_i) = f_i, \ i = 0, 1, \ldots, N+1, \ Du(x_i) = f_i{}^1, \ i = 0, N+1\}.$$

Theorem 8.6 $p_s f$ is the unique solution to the problem: find $u \in U_f$ to minimize $\|D^2 u\|_2{}^2$.

Proof This closely follows that of Theorem 8.2: the details are left to exercise 23.

□

This theorem shows that the cubic spline is the interpolating function of minimum least squares curvature. If the interpolation data corresponds to function values $f_i = f(x_i)$, $i = 0, 1, \ldots, N+1$ and derivatives $f_i{}^1 = Df(x_i)$, $i = 0, N+1$, we can regard the situation as that of the draughtsman's flexible rod referred to earlier being forced to pass through the points $(x_i, f(x_i))$, $i = 0, 1, \ldots, N+1$ and being clamped at the end points. If $f(x)$ is sufficiently smooth, then error bounds similar to those derived in the Hermite case can again be obtained, and these show that the cubic spline interpolant is fourth order accurate. We merely state the following results without proof; for detailed proofs see Hall (1968) and Schultz (1973).

Theorem 8.7 If $f(x) \in C^4[a, b]$, then

$$\|f - p_s f\|_2 \leqslant 4\pi^{-4} h^4 \|D^4 f\|_2$$

$$\|f - p_s f\|_\infty \leqslant \frac{5}{384} h^4 \|D^4 f\|_\infty.$$

Lower order bounds for the derivatives of the error may also be obtained. As in the Hermite case, the order of accuracy is then $4 - k$, where k is the order of derivative. In addition, analogues of these results are available in two dimensions for the bicubic spline interpolant introduced in exercise 20 (see Schultz, 1969, 1973, and Carlson and Hall, 1973, for details).

Exercises

22. For the differential equation of exercise 8, obtain the finite element matrix appropriate to the space $S^{(2)}(\Delta)$, using the B-spline basis.

23. Prove Theorem 8.6.

24. Let V be the set of functions

$$V = \{v(x) : v(x) \in C[a, b], v(x) \in C^1(x_i, x_{i+1}), i = 0, 1, \ldots, N\}.$$

For some constant $\alpha \geqslant 0$ and some $f(x)$ defined in $[a, b]$, show that the function $v(x)$ which solves the problem

$$\text{find } v \in V \text{ to minimize } \left\{ \int_a^b [Dv]^2 \, dx + \alpha \sum_{i=0}^{N-1} [v(x_i) - f(x_i)]^2 \right\}$$

is an element of $S^{(1)}(\Delta)$. If $v(x)$ is expressed in terms of B-splines, show that the coefficients satisfy a tridiagonal system of $N + 2$ linear equations.

In the case where $x_i = ih$, $i = 0, 1, \ldots, N + 1$, $(N+1)h = 1$, and $\alpha = h$, consider the relationship between the above problem and the solution of the ordinary differential equation

$$D^2 y + y = f \qquad 0 \leqslant x \leqslant 1$$

where $D^2 y$ at x_i is replaced by the central difference approximation

$$(y_{i+1} - 2y_i + y_{i-1})/h^2.$$

25. A *monospline* of degree $n + 1$ with knots defined by Δ can be written

$$k(x) = \frac{x^{n+1}}{(n+1)!} + g(x)$$

where $g(x) \in S^{(n)}(\Delta)$. Prove that if the quadrature formula

$$\int_a^b f(x) \, dx \approx \sum_{i=0}^{N+1} w_i f(x_i)$$

is exact when $f(x)$ is a polynomial of degree $n \leqslant N + 1$, then the remainder $R(f)$ can be written

$$R(f) = \int_a^b D^{(n+1)} f(x) k(x) \, dx$$

if $f(x) \in C^{n+1}[a, b]$, where $k(x)$ is a monospline of degree $(n + 1)$ with knots defined by Δ. Obtain an explicit expression for $k(x)$.

If $n < N + 1$, then the degrees of freedom can be removed by making $\int_a^b (k(x))^2 \, dx$ a minimum (giving a 'best' formula). Derive a system of linear equations whose solution will give the unknowns $\{w_i\}$ in this case, and hence obtain the formula

$$\int_0^2 f(x) \, dx \approx \tfrac{1}{8}[3f(0) + 10f(1) + 3f(2)]$$

for the case $n = 1$.

The definition of $k(x)$ can be generalized if we permit the continuity of $g(x)$ at the knots to be reduced so that $g(x) \in C^{n-r}[a, b]$ $(1 \leqslant r \leqslant n + 1)$. Indicate the form of the resulting quadrature formula which may be derived in this case, and obtain an expression for $k(x)$, with $r = n$ and integer knots, which will give the formula of this type which is best for the interval $[0, N + 1]$. (See, for example, Schoenberg, 1965; Karlin, 1971).

8.6 Methods for approximating by spline functions

An important application of piecewise polynomials is to the approximation of functions defined *implicitly*, for example by differential or integral equations. The structure and analysis of most of the popular methods (for example Rayleigh–Ritz–Galerkin or finite element methods) depends substantially on the particular nature of the original problem whose solution is required, and therefore we do not consider this further. We turn instead to the use of piecewise polynomials to approximate a function defined *explicitly*, either as a continuous function or as a discrete set of data (which may be obtained by sampling a continuous function at a finite set of points). The fitting of spline functions has attracted most attention, and we will restrict consideration to this particular class of approximating functions; in certain degenerate cases, however, piecewise polynomials not in $S^{(n)}(\Delta)$ may be permitted.

We assume at present that the knot positions are preselected, so that the problem is a linear approximation problem, and the analyses and techniques described in the previous chapters are available. Only the L_2 and L_∞ norms have attracted any real attention. The L_2 norm has been popular, one reason for this being the special band structure of the matrix of the normal equations, resulting from the compact support of the B-splines (exercise 26). For the discrete least squares problem, the stable variant described in Chapter 4 which uses orthogonal factorization can be efficiently implemented. For example, Hayes and Halliday

(1974) give a method for the cubic case which takes full advantage of the structure of the matrix A of the linear system (see also Reid, 1967; Cox and Hayes, 1973), and readily generalizes to more than one variable.

The theoretical analysis of the L_∞ problem is complicated by the fact that approximations are not being sought from a Haar subspace of $C[a, b]$. However, the fact that the set of functions (see exercise 13)

$$\{1, x, \ldots, x^n, (x - x_1)_+{}^n, \ldots, (x - x_N)_+{}^n\}$$

forms a weak Chebyshev set in $[a, b]$ (exercise 19 of chapter 3) does in fact permit best approximations to be characterized by alternation-type conditions. We state, without proof, the following result (see, for example, Rice 1967; Schumaker 1969).

Theorem 8.8 Let $f(x) \in C[a, b]$. Then
  (i) $g(x) \in S^{(n)}(\Delta)$ is a best approximation to $f(x)$ if and only if $f(x) - g(x)$ alternates at least $n + k + 1$ times on some interval $[x_i, x_{i+k+1}]$, $0 \leqslant i \leqslant i + k + 1 \leqslant N + 1$,
  (ii) there exists $g(x) \in S^{(n)}(\Delta)$ such that $f(x) - g(x)$ alternates at least $N + n + 1$ times on $[a, b]$.

The absence of a Chebyshev set means of course that uniqueness of the best approximation is not guaranteed, and in fact there must be at least one $f(x) \in C[a, b]$ with more than one best approximation (Theorem 3.5). An example is given in exercise 27. The following example, due to Schumaker (1969), also illustrates that $N + n + 1$ alternations are not necessary for a best approximation.

Example Let $[a, b] = [-1, 1]$, $n = N = 1$, $x_1 = 0$ and
$$f(x) = \begin{cases} 1 & -1 \leqslant x \leqslant -\tfrac{1}{2} \\ -2x & -\tfrac{1}{2} \leqslant x \leqslant 0 \\ 2x & 0 \leqslant x \leqslant 1. \end{cases}$$

Theorem 8.8 shows that $g(x) = \tfrac{1}{4} - x + cx_+$ is a best approximation for all c satisfying $\tfrac{3}{2} \leqslant c \leqslant 3$. Only the choice $c = \tfrac{5}{2}$ gives $N + n + 1 = 3$ alternations.

It follows from Theorem 8.8 that, of the methods described in Chapter 3, only the more general are available for best L_∞ approximation by splines. A recent method which applies specifically to approximation by a linear combination of functions forming a weak Chebyshev set has been given by Karon (1978). The method, which is of descent type, therefore applies in particular to spline approximation; numerical results are promised in Karon and Starner (to appear). A linear programming based method for L_∞ approximation on a discrete subset of $[a, b]$ is given by Schultz (1972).

The intrinsic simplicity of interpolation has resulted in most of the work on approximation by splines being concerned with this particular criterion. For odd degree splines, it is convenient to choose the knots to coincide with points at which data values are given and to fit the spline according to Theorem 8.4, where the values f_i^k are assumed to be appropriate kth derivatives of the function $f(x)$ being approximated. This additional derivative information may not always be available, however, and Hayes (1970) suggests the finite difference replacement

$$g(y_t) - f(y_t) = g(z_t) - f(z_t) \qquad t = 1, N+1,$$

when $n = 3$, where

$$y_t = \tfrac{1}{2}(x_{t-1} + x_t), \ z_t = \tfrac{1}{2}(x_t + x_{t+1}) \qquad t = 1, N+1.$$

Another way of removing the additional degrees of freedom is through the use of *natural splines*.

Definition 8.5 A spline $g(x)$ of odd degree $n = 2m - 1$ is called a natural spline if it is given in each of the two intervals $[x_0, x_1]$ and $[x_N, x_{N+1}]$ by a polynomial of degree $m - 1$ (in general, not the same polynomial in both intervals).

It is straightforward to show (exercise 28) that $g(x) \in S^{(n)}(\Delta)$ is a natural spline if and only if it is of the form

$$g(x) = \sum_{i=0}^{m-1} a_i x^i + \sum_{i=0}^{N} c_i (x - x_i)_+^n \qquad (8.21)$$

with

$$\sum_{i=1}^{N} c_i x_i^r = 0 \qquad r = 0, 1, \ldots, m-1.$$

Provided that $m \leqslant N$, there is a unique natural spline satisfying the interpolation conditions

$$g(x_i) = f_i \qquad i = 1, 2, \ldots, N. \qquad (8.22)$$

The proof of this result is left to exercise 30. An important property of natural splines is that they solve a certain best approximation problem. Define the set V_f by

$$V_f = \{v(x) : v(x) \in C^{m-1}[a, b], \ v(x) \in C^m(x_i, x_{i+1}), \ i = 0, 1, \ldots, N,$$
$$v(x_i) = f_i, \ i = 1, 2, \ldots, N\}.$$

Theorem 8.9 The natural spline $g(x) \in S^{(n)}(\Delta)$ satisfying (8.22) is the unique solution to the problem: find $v \in V_f$ to minimize $\| D^m v \|_2^2$.

For a proof of this general result, see de Boor (1963). The proof for the case $m = 2$ is similar to that of Theorem 8.2 and is left as an exercise.

A computational scheme for interpolation by natural splines is given by Lyche and Schumaker (1973). One disadvantage of the use of these splines as interpolants is that there is reduced accuracy at the ends of the interval. For example, in the cubic case if $f(x) \in C^4[a, b]$ is being approximated in $[x_1, x_N]$, then unless $D^2 f(x_1) = D^2 f(x_N) = 0$ the accuracy of the approximation is $O(h^2)$, where h is the maximum knot spacing. However, for any subinterval $[c, d] \subset [x_1, x_N]$, the error is $O(h^4)$, (See Kershaw, 1971; Hall, 1973).

More general spline interpolation schemes are of course possible. For example, if the total number of data values is t, then we may choose N and n so that $N + n + 1 = t$, and only choose some of the data points as knots. For such interpolating functions to exist, some restriction on the knot choice is necessary: Schoenberg and Whitney (1953) show that $g(x) \in S^{(n)}(\Delta)$ interpolating to the data (y_i, f_i), $i = 1, 2, \ldots, t$ exists if and only if the knots satisfy

$$y_i < x_i < y_{i+n+1} \qquad i = 1, 2, \ldots, N.$$

(See, for example, Rice, 1969, for a proof of this.)

Example $n = 1, t = 5, N = 3, y_i = i, i = 1, 2, \ldots, 5$.

If $\Delta = \{2, 3, 4\}$, then the conditions are satisfied; if $\Delta = \{1, 2, 3\}$ they are not. The existence question is easily answered directly by calculation of the determinants of the matrices arising from the representation of exercise 13, giving values 1 and 0 respectively.

An algorithm for this interpolation problem is given by Cox (1975).

As already indicated, considerable improvement in approximating capacity is possible if the knot positions are allowed to be variables. Illustrations of this improvement have been given by, for example, Burchard (1974). Returning to the problem of best approximation by splines with free knots, the first difficulty is that the existence of a best approximation is no longer guaranteed. The following example is due to Schumaker (1969).

Example For $n > 1$, let $f(x) = nx_+^{n-1}$, and let a best approximation be sought in $[-1, 1]$ by splines of degree n with 2 free knots. Clearly $f(x)$ is not a member of this class. However,

$$g_k(x) = \frac{k}{2}\left(x + \frac{1}{k}\right)_+^n - \frac{k}{2}\left(x - \frac{1}{k}\right)_+^n$$

is a sequence of appropriate splines with

$$\|g_k(x) - f(x)\|_\infty \to 0 \quad \text{as} \quad k \to \infty.$$

Thus no best L_∞ approximation exists.

This example illustrates that it is necessary to permit the occurrence of *multiple knots*: a spline $g(x)$ of degree n is said to have a knot of multiplicity k_i at the point x_i provided that there exists an open neighbourhood N_i of x_i such that $g(x) \in C^{n-k_i}(N_i)$, $g(x) \notin C^{n-k_i+1}(N_i)$. Such a function $g(x)$ is often referred to as an *extended spline*. Schumaker (1968) shows that the existence of a best L_∞

approximation can be guaranteed from the space $S_N^{(n)}$ of spline functions of degree n with N knots in the interval $[a, b]$, allowing multiple knots up to multiplicity $(n + 1)$. He also shows that $g(x) \in S_N^{(n)}$ if and only if $g(x)$ has the form

$$g(x) = \sum_{i=0}^{n} a_i x^i + \sum_{i=1}^{r} \sum_{j=1}^{k_i} c_{ij}(x - x_i)_+^{n-j+1}$$

with $\sum_{i=1}^{r} k_i = N$. Such a spline may also be expressed in terms of suitably extended B-splines (see exercise 36).

Alternation type characterization results are also available for the L_∞ free-knot problem. Necessary conditions and sufficient conditions are given by Braess (1971), who shows that the gap between them cannot be bridged: this gap is a consequence of the fact that best approximations are not always unique.

Because of the numerical difficulties involved in approximating by splines with free knots, most methods have achieved some degree of variability of knot positions by adaptive knot *addition* and *subtraction*. Cox (1977) suggests that a sensible method, which should lead to satisfactory approximations, can be based on the following iteration scheme:

(i) preselect a set of knots by using any available information about the function to be approximated
(ii) determine an approximating spline having these knots
(iii) examine the error of the approximation: in regions where the approximation is inadequate, introduce additional knots; remove knots where there appear to be too many; adjust other current knot positions if this seems appropriate; return to (ii).

An algorithm based on this sort of approach is given by Powell (1970) for fitting a cubic spline to discrete data. The underlying criterion of (ii) above is the least squares norm, but a restriction is placed on the size of the discontinuity in the third derivative at the knots. When the discrete data comes from a differentiable function, a related method is given by Curtis (1970), and a Fortran program provided by Powell (1972). Other methods have been given, for example by de Boor and Rice (1968), and by de Boor (1973). The former determines a spline with as few knots as possible whose least squares error is less than a prescribed tolerance; the latter attempts to use nth derivative function information to obtain good knot positions.

A recent method which attempts to find a best L_2 approximation by a spline whose knots are genuinely free parameters is given by Jupp (1978). The method introduces a transformation of knot variables in order to reduce the numerical difficulties involved, and provided that a good initial estimate of the knot positions is available, appears to work well. An L_∞ descent method is given by Karon (1978).

Exercises

26. Let $f(x) \in C[a, b]$. Show that the problem of approximating to $f(x)$, using the L_2 norm, by elements of $S^{(1)}(\Delta)$ in B-spline representation results in the normal equations being a tridiagonal system of $N+2$ linear equations with positive definite coefficient matrix. What is the general form of the matrix which results if approximations are sought from $S^{(n)}(\Delta)$?

27. Let $[a, b] = [0, 1]$ and let $N = 1$ and $x_1 = 0.5$. Demonstrate non-uniqueness of the best L_∞ approximation by elements of $S^{(1)}(\Delta)$ by taking $f(x) = \exp\{k(x-1)\}$, where $k \gg 1$.

28. Prove that $g(x) \in S^{(n)}(\Delta)$ is a natural spline if and only if it satisfies the conditions (8.21).

29. Let $n = 2m - 1$ and let $g(x) \in S^{(n)}(\Delta)$ be a natural spline. Let $f(x) \in C^{m-1}[a, b]$ with $f(x) \in C^m(x_i, x_{i+1})$, $i = 0, 1, \ldots, N$, and let $f(x_j) = 0$, $j = 1, 2, \ldots, N$. Prove that

$$\int_a^b D^m f(x) D^m g(x) \mathrm{d}x = 0.$$

30. If $m \le N$, prove that there exists a unique natural spline $g(x) \in S^{(n)}(\Delta)$ satisfying the interpolation conditions (8.22) (c.f. Theorem 8.4).

31. For the case $m = 3$, prove Theorem 8.9.

32. Let Δ be an equispaced set of points in $[a, b]$ and let $g(x) \in S^{(3)}(\Delta)$ interpolate $f(x) \in C^1[a, b]$ at x_i, $i = 0, 1, \ldots, N+1$. Let the remaining degrees of freedom on the spline be used up by satisfying

$$Dg(x_0) = \alpha$$
$$D^2 g(x_0) = \beta.$$

Use Taylor series expansion to show that the derivative of the spline defined in this way can be evaluated on Δ by means of the recurrence relations

$$hDg(x_{i+1}) = 3(f(x_{i+1}) - f(x_i)) - 2hDg(x_i) - \tfrac{1}{2}h^2 D^2 g(x_i)$$
$$h^2 D^2 g(x_{i+1}) = 6(f(x_{i+1}) - f(x_i)) - 6hDg(x_i) - 2h^2 D^2 g(x_i),$$

$i = 1, 2, \ldots, N$, where $h = (b-a)/(N+1)$.

By considering the behaviour of $Dg(x_{N+1})$, $D^2 g(x_{N+1})$ as $N \to \infty$, show that if $\alpha = Df(x_0)$, $\beta = D^2 f(x_0)$, then this method of approximation will not cause $Dg(x)$ to converge to $Df(x)$ for all $x \in [a, b]$. Illustrate this for the case $f(x) = \exp(-x)$ in $[0, 1]$, taking $N = 9$ and $N = 19$.

33. Let $g(x)$ be the interpolative cubic spline defined in the previous exercise, but let the additional degrees of freedom be removed by setting

$$Dg(x_0) = \alpha$$
$$Dg(x_{N+1}) = \gamma.$$

Show that the first and second derivative values of this spline may be obtained by using the recurrence relations of the previous exercise for two trial values of $D^2 g(x_0)$, and using linear interpolation. When $\alpha = Df(x_0)$, $\gamma = Df(x_{N+1})$, illustrate the situation in this case for $f(x) = \exp(-x)$ in $[0, 1]$, taking $N = 9$ and $N = 19$, and compare with the results of the previous exercise.

34. Consider the problem of approximating to $f(x) \in C[0, 1]$ by elements of $S^{(1)}(\Delta)$ when the knots $x_1, x_2, \ldots, x_N$ are allowed to become additional variables of the problem. Show that the L_2 norm of the error is minimized when $g(x) \in S^{(1)}(\Delta)$ is such that on any subinterval (x_i, x_{i+1}), $g(x)$ gives the best L_2 approximation to $f(x)$ by a straight line on that subinterval. What is the effect of coalescing knots?

35. Use the result of exercise 34 to determine the best L_2 approximation from $S^{(1)}(\Delta)$, where $N = 1$ and the knot x_1 is assumed variable, to $f(x) = x^2(1 - x)$.

36. B-splines having multiple knots to order $n + 1$ may be defined by (8.15), the relation (8.18) remaining valid (de Boor, 1972). For example, if we take

$$x_{-n} = x_{-n+1} = \ldots = x_0, \quad x_{N+1} = x_{N+2} = \ldots = x_{N+n+1},$$

then the full set of B-splines is nonzero only in $[a, b]$. If $g(x)$ is given by (8.17), show that

$$\int_a^b g(x)\, dx = \frac{1}{(n+1)} \sum_{i=1}^{N+n+1} a_i \qquad \text{(Cox, 1975)}.$$

For more on splines, see de Boor (1977, 1978).

9

Chebyshev approximation by nonlinear families

9.1 Introduction

To this point, we have been almost exclusively concerned with approximation where the space of approximating functions is linear. It is clear that this forces consideration of an extremely small (yet very important) subclass of approximating functions, and therefore excludes many forms which may be more appropriate in a given situation. One reason for the emphasis on linearity is the rather obvious one that from both the theoretical and practical point of view there is considerable simplification in the required analysis. For example, we have seen that the removal of linearity means that the subspace of approximations is no longer finite dimensional, and an immediate consequence of this is that compactness, and therefore existence of a best approximation, can no longer be automatically guaranteed. Even if it is felt that this is a theoretical point, which may be ignored by the practical scientist, there still remain the potentially difficult tasks of characterizing these nonlinear best approximations and providing good numerical methods for their calculation.

Work on nonlinear approximation has been largely concerned with the Chebyshev norm, and with the treatment of rather special subclasses of problem whose solutions exhibit features similar to those occurring in the continuous linear case. In particular, the important class of rational approximating functions has been extensively studied, and most of this chapter is concerned with a detailed study of this class. There are two main reasons for the importance of such functions: firstly, they have good approximating properties, and can give considerable improvement over, for example, ordinary polynomials; and secondly, the theory of approximation in the Chebyshev norm is extremely well developed, and efficient numerical methods are available. We begin, however, by widening the discussion to more general nonlinear Chebyshev approximation problems, and give an outline of analysis which is available in this case. The following section may be omitted without loss of continuity.

9.2 Nonlinear approximation on an interval

The formulation of a general characterization theory for nonlinear Chebyshev approximation is complicated by the problem of dealing with the gap between the conditions which are necessary and those which are sufficient for a best approximation. If sufficient conditions, for example, are derived by a natural generalization of those appropriate in the linear case, then the class of approximating functions has to be restricted in some way in order that the conditions may also be necessary, and vice versa. The point at which these theories naturally converge, so that conditions which are both necessary and sufficient become available, may conveniently be interpreted in terms of an alternation-type condition, analogous to that which applies in the linear Chebyshev set case. In order to arrive at such a result directly, the concept of *unisolvency* was introduced by Motzkin in 1949. Let $\phi(x, \mathbf{a})$ be a nonlinear mapping of R^n into $C[a, b]$; then given any $\mathbf{d} \in R^n$ and any n distinct points x_i, $i = 1, 2, \ldots, n$, in $[a, b]$, the family of approximating functions

$$\{\phi(x, \mathbf{a}) \in C[a, b], \mathbf{a} \in R^n\}$$

is unisolvent if there exists a unique vector $\mathbf{a} \in R^n$ such that

$$\phi(x_i, \mathbf{a}) = d_i \qquad i = 1, 2, \ldots, n.$$

This particular generalization of the Chebyshev set property in the linear case can be shown to lead to the existence of a unique best approximation to $f(x) \in C[a, b]$ which is characterized by $(n + 1)$ points of alternation of the error curve (Tornheim, 1950; see also Rice, 1964a). Unfortunately, this is an extremely restrictive property, possessed by only a small number of nonlinear approximating functions. A rather more general property was suggested by Rice (1960) called *varisolvency*, which led to the best approximation being characterized by an alternating set, the number of points in which was allowed to vary.

A related theory for nonlinear Chebyshev approximation on an interval was established by Meinardus and Schwedt (1964), starting from generalizations of the Kolmogoroff criterion (exercise 2 of Chapter 3); for given $\mathbf{a} \in R^n$, we denote by $\overline{E}(\mathbf{a})$ the set

$$\overline{E}(\mathbf{a}) = \{x : x \in [a, b], |f(x) - \phi(x, \mathbf{a})| = \|f - \phi\|\}$$

for given $f(x) \in C[a, b]$.

Theorem 9.1 If

$$\min_{x \in \overline{E}(\mathbf{a})} (f(x) - \phi(x, \mathbf{a}))(\phi(x, \mathbf{c}) - \phi(x, \mathbf{a})) \leqslant 0 \qquad (9.1)$$

for all $\mathbf{c} \in R^n$, then $\phi(x, \mathbf{a})$ is a best approximation to $f(x)$.

Proof Let the condition be satisfied, but $\phi(x, \mathbf{a})$ not be a best approximation. Then there exists $\mathbf{c} \in R^n$ such that

$$\| f(x) - \phi(x, \mathbf{c}) \| < \| f(x) - \phi(x, \mathbf{a}) \|.$$

Thus for all $x \in \overline{E}$ (**a**)

$$(f(x) - \phi(x, \mathbf{a}))(\phi(x, \mathbf{c}) - \phi(x, \mathbf{a})) > 0$$

which gives a contradiction and proves the result. □

Now let $\phi(x, \mathbf{a})$ be a best approximation, and suppose that (9.1) is not satisfied for some $\mathbf{c} \in R^n$. Then for any $x \in \overline{E}(\mathbf{a})$

$$(f(x) - \phi(x, \mathbf{a}))(\phi(x, \mathbf{c}) - \phi(x, \mathbf{a})) > 0$$

and so using exercise 3 of Chapter 3

$$\| f(x) - (1 - \gamma)\phi(x, \mathbf{a}) - \gamma\phi(x, \mathbf{c}) \| < \| f(x) - \phi(x, \mathbf{a}) \|$$

for $\gamma > 0$ sufficiently small. This inequality shows that the condition of Theorem 9.1 is *necessary* for a best approximation if there exists $\mathbf{d}(\gamma) \in R^n$ such that

$$\phi(x, \mathbf{d}(\gamma)) = (1 - \gamma)\phi(x, \mathbf{a}) + \gamma\phi(x, \mathbf{c}). \tag{9.2}$$

In fact, necessity may be proved under the rather weaker assumption that the family of approximating functions is *asymptotically convex*: for any $\mathbf{a}$, $\mathbf{c} \in R^n$ and any real γ, $0 \leqslant \gamma \leqslant 1$, there exists $\mathbf{d}(\gamma) \in R^n$ and a continuous function $g(x, \gamma)$ defined on $[a, b] \times [0, 1]$ with $g(x, 0) > 0$ such that

$$\| \phi(x, \mathbf{d}(\gamma)) - (1 - \gamma g(x, \gamma))\phi(x, \mathbf{a}) - \gamma g(x, \gamma)\phi(x, \mathbf{c}) \| = o(\gamma) \tag{9.3}$$

(for example, Meinardus, 1967), where $A = o(\gamma)$ means that $\lim_{\gamma \to 0} A/\gamma = 0$.

Now assume that the partial derivatives of the function $\phi(x, \mathbf{a})$ with respect to the components of $\mathbf{a}$ exist and are continuous. Then necessary conditions for a best approximation can be given as shown in the following theorem (due to Chebyshev), which again may be regarded as a (partial) generalization of the Kolmogoroff criterion.

Theorem 9.2 Let $\phi(x, \mathbf{a})$ be a best approximation to $f(x) \in C[a, b]$. Then

$$\min_{x \in \overline{E}(\mathbf{a})} (f(x) - \phi(x, \mathbf{a})) \sum_{i=1}^{n} c_i \frac{\partial \phi(x, \mathbf{a})}{\partial a_i} \leqslant 0$$

for all $\mathbf{c} \in R^n$.

Proof Let $\phi(x, \mathbf{a})$ be a best approximation, but assume that there exists $\mathbf{c} \in R^n$ such that for all $x \in \overline{E}(\mathbf{a})$

$$(f(x) - \phi(x, \mathbf{a})) \sum_{i=1}^{n} c_i \frac{\partial \phi(x, \mathbf{a})}{\partial a_i} > 0.$$

Then for $\gamma > 0$, any $x \in \overline{E}(\mathbf{a})$,

$$(f(x) - \phi(x, \mathbf{a}))(\phi(x, \mathbf{a} + \gamma \mathbf{c}) - \phi(x, \mathbf{a}) + o(\gamma)) > 0.$$

It follows that for $\gamma > 0$ sufficiently small

$$\| f(x) - \phi(x, \mathbf{a} + \gamma \mathbf{c}) \| < \| f(x) - \phi(x, \mathbf{a}) \|$$

(exercise 2), which gives a contradiction, and proves the result. $\qquad \square$

Now let the conditions of Theorem 9.2 be satisfied, but suppose that $\phi(x, \mathbf{a})$ is not a best approximation. Then there exists $\mathbf{c} \in R^n$ such that

$$|f(x) - \phi(x, \mathbf{a} + \mathbf{c})| < |f(x) - \phi(x, \mathbf{a})| \qquad x \in \overline{E}(\mathbf{a}). \tag{9.4}$$

In order for this to lead to a contradiction, and thus force the conditions of Theorem 9.2 to be sufficient as well as necessary, Krabs (1967) suggests requiring the following *representation condition*: for any members $\phi(x, \mathbf{a})$, $\phi(x, \mathbf{b})$ of the family of approximating functions, there exists a real vector $\boldsymbol{\delta} \equiv \boldsymbol{\delta}(\mathbf{a}, \mathbf{b}) \in R^n$ and a function $K \equiv K(\mathbf{a}, \mathbf{b}, x)$ positive on $[a, b]$ such that

$$\phi(x, \mathbf{a}) - \phi(x, \mathbf{b}) = K \sum_{i=1}^{n} \delta_i \frac{\partial \phi(x, \mathbf{a})}{\partial a_i}.$$

In this case, (9.4) can be written

$$\left| f(x) - \phi(x, \mathbf{a}) + K \sum_{i=1}^{n} \delta_i \frac{\partial \phi(x, \mathbf{a})}{\partial a_i} \right| < |f(x) - \phi(x, \mathbf{a})| \qquad x \in \overline{E}(\mathbf{a}),$$

so that the conditions of Theorem 9.2 are contradicted.

Let us now denote by $W(\mathbf{a})$ the linear space spanned by the n partial derivatives of $\phi(x, \mathbf{a})$, and let $d(\mathbf{a}) (\leqslant n)$ denote its dimension. Then Meinardus and Schwedt (1964) give the following alternation-type characterization theorem.

Theorem 9.3 Assume that
  (i) for every $\mathbf{a} \in R^n$, $W(\mathbf{a})$ is a Haar subspace of dimension $d(\mathbf{a})$,
  (ii) for every $\mathbf{a}, \mathbf{c} \in R^n$, the function $\phi(x, \mathbf{a}) - \phi(x, \mathbf{c})$ either possesses at most $d(\mathbf{a}) - 1$ zeros in $[a, b]$ or vanishes identically.
Then $\phi(x, \mathbf{a})$ is a best approximation to $f(x) \in C[a, b]$ if and only if there exists an alternating set of $d(\mathbf{a}) + 1$ points in $[a, b]$.

Proof *Sufficiency* Let the conditions be satisfied, but $\phi(x, \mathbf{a})$ not be a best approximation. Then for some $\mathbf{c} \in R^n$

$$\| f(x) - \phi(x, \mathbf{c}) \| < \| f(x) - \phi(x, \mathbf{a}) \|$$

so that $\phi(x, \mathbf{a}) - \phi(x, \mathbf{c})$ alternates in sign on the alternating set. This contradicts assumption (ii) and proves the result.
Necessity The conditions of Theorem 9.2 give, by Theorem 1.5,

$$0 \in \text{conv}\{\psi(x)\,\text{sign}\,(f(x) - \phi(x, \mathbf{a})), x \in \overline{E}(\mathbf{a})\}$$

where $\psi(x) \in R^{d(\mathbf{a})}$ and has ith component $\psi_i(x, \mathbf{a})$, $i = 1, 2, \ldots, d(\mathbf{a})$, where $\{\psi_i(x, \mathbf{a})\}$ is a basis for $W(\mathbf{a})$. Then, by the Haar condition assumption, there exist $t = d(\mathbf{a}) + 1$ points $x_1, x_2, \ldots, x_t$ in $\overline{E}(\mathbf{a})$ and $\lambda \in R^t$ such that

$$A^T \lambda = 0$$

with $\lambda_i \operatorname{sign}(f(x_i) - \phi(x_i, \mathbf{a})) > 0, i = 1, 2, \ldots, t, \sum_{i=1}^{t} |\lambda_i| = 1$, where A is the $t \times (t - 1)$ matrix with (i,j) element $\psi_i(x_i, \mathbf{a})$, $i = 1, 2, \ldots, t, j = 1, 2, \ldots, t - 1$. Finally, the components of λ must alternate in sign, by Lemma 3.1, and the required result follows. $\qquad \square$

Theorem 9.4 If assumptions (i) and (ii) of Theorem 9.3 hold, there is at most one best approximation to $f(x) \in C[a, b]$.

Proof Let $\phi(x, \mathbf{a})$ and $\phi(x, \mathbf{c})$ both be best approximations. Then there exist points x_i, $i = 1, 2, \ldots, d(\mathbf{a}) + 1$, in $\overline{E}(\mathbf{a})$ such that (without loss of generality)

$$f(x_i) - \phi(x_i, \mathbf{a}) = (-1)^i \| f(x) - \phi(x, \mathbf{a}) \| \qquad i = 1, 2, \ldots, d(\mathbf{a}) + 1,$$

so that

$$(-1)^i (\phi(x_i, \mathbf{c}) - \phi(x_i, \mathbf{a})) \geq 0 \qquad i = 1, 2, \ldots, d(\mathbf{a}) + 1.$$

If $\phi(x_i, \mathbf{c}) = \phi(x_i, \mathbf{a})$, $i = 1, 2, \ldots, d(\mathbf{a}) + 1$, then $\phi(x, \mathbf{c}) = \phi(x, \mathbf{a})$ by assumption (ii). Thus assume that

$$\operatorname{sign}(\phi(x_k, \mathbf{c}) - \phi(x_k, \mathbf{a})) = (-1)^k$$

for some $k \in [1, d(\mathbf{a}) + 1]$. Now by assumption (i), there exists a unique vector $\mathbf{b} \in R^n$ such that

$$\sum_{j=1}^{d(\mathbf{a})} b_j \psi_j(x_i, \mathbf{c}) = (-1)^i \qquad i = 1, 2, \ldots, d(\mathbf{a}) + 1, i \neq k.$$

Further,

$$\phi(x, \mathbf{c} + \gamma \mathbf{b}) - \phi(x, \mathbf{a}) = \phi(x, \mathbf{c}) - \phi(x, \mathbf{a}) + \gamma \sum_{j=1}^{d(\mathbf{a})} b_j \psi_j(x, \mathbf{c}) + O(\gamma)$$

so that for $\gamma > 0$ sufficiently small,

$$\operatorname{sign}(\phi(x_i, \mathbf{c} + \gamma \mathbf{b}) - \phi(x_i, \mathbf{a})) = (-1)^i \qquad i = 1, 2, \ldots, d(\mathbf{a}) + 1.$$

Thus the continuous function $\phi(x, \mathbf{c} + \gamma \mathbf{b}) - \phi(x, \mathbf{a})$ has $d(\mathbf{a})$ zeros in $[a, b]$ and so must vanish identically. By continuity, this must hold on letting $\gamma \to 0$ and the required result is obtained. $\qquad \square$

The way in which varisolvent functions are defined has the effect of giving precisely equivalent characterization and uniqueness results (Rice, 1960). In fact there is rather more generality in this class, as the existence of partial derivatives is not required; however, from a practical point of view, the availability of these derivatives is extremely important. Braess (1974b) demonstrates the precise relationship between the various assumptions. (The assumption (i) of Theorem 9.3 is usually referred to as the local Haar condition, and assumption (ii) as the global Haar condition.)

In attempting to define a general class of nonlinear approximating functions having properties which could be exploited in satisfying conditions equivalent to those of Theorem 9.3, Hobby and Rice (1967) introduced the γ-polynomials

$$\phi(x, \mathbf{a}) = \sum_{i=1}^{n} a_i \gamma(a_{i+n}, x)$$

where γ is a continuous function of its parameters. A more general class of functions consisting of the closure of the γ-polynomials has also been studied. Although allowing the existence of a best approximation to be guaranteed, the characterization of best approximations from this wider class is rather more complicated. A general analysis of functions of this type is beyond the scope of this book—the interested reader is referred to Rice (1969), Braess (1973, 1974a), and the references which are to be found there. We will merely look at the important special case of approximation by sums of exponentials, where we have $\gamma(t, x) = e^{tx}$, so that

$$\phi(x, \mathbf{a}) = \sum_{i=1}^{n} a_i e^{t_i x}, \tag{9.5}$$

where, for convenience, we have renamed a_{i+n} as t_i, $i = 1, 2, \ldots, n$. Assume that $a_1, a_2, \ldots, a_k$ are nonzero, $a_i = 0$, $i = k+1, \ldots, n$, and $t_1 > t_2 > \ldots > t_k$, with the remaining values of t_i being arbitrary, but such that $t_i \neq t_j$, $i \neq j$. Then the gradient vector is

$$\mathbf{g}(x, \mathbf{a}) = (e^{t_1 x}, \ldots, e^{t_n x}, a_1 x e^{t_1 x}, \ldots, a_n x e^{t_n x})^T,$$

which has $n + k$ nonzero components, and thus $W(\mathbf{a})$ has dimension $d(\mathbf{a}) = n + k$.

Theorem 9.5 $\phi(x, \mathbf{a})$ is a best approximation to $f(x) \in C[a, b]$ from the class (9.5) if and only if there exists an alternating set of $n + k + 1$ points in $[a, b]$. At most one best approximation exists.

Proof We will show that the conditions of Theorem 9.3 are satisfied with $d(\mathbf{a}) = n + k$, when the results will follow from that theorem and Theorem 9.4. Let

$$P(x, r) = \sum_{i=1}^{r} (b_i + c_i x) e^{t_i x}. \tag{9.6}$$

We begin by proving that $P(x, r)$ has at most $(2r - 1)$ zeros. The result is obvious for $r = 1$, so we proceed by induction, and assume that $P(x, s)$ has at most $(2s - 1)$

zeros. Then

$$P(x, s+1) = \sum_{i=1}^{s+1} (b_i + c_i x)e^{t_i x}$$

$$= (b_1 + c_1 x + \sum_{i=2}^{s+1} (b_i + c_i x)e^{(t_i - t_1)x})e^{t_1 x}$$

and the zeros of $P(x, s+1)$ are those of the coefficient of $e^{t_1 x}$. If this has more than $(2s + 1)$ zeros in $[a, b]$, then the second derivative with respect to x has, by Rolle's theorem, more than $(2s - 1)$ zeros in $[a, b]$. The form of the second derivative shows that this contradicts the induction hypothesis. Thus $P(x, s+1)$ has at most $(2s + 1)$ zeros, and the result that $P(x, r)$ has at most $(2r - 1)$ zeros is proved.

Now assumption (i) of Theorem 9.3 requires that any element

$$\sum_{i=1}^{n} b_i e^{t_1 x} + x \sum_{i=1}^{k} c_i a_i e^{t_i x}$$

has at most $n + k - 1$ zeros, and this follows immediately. Assumption (ii) is satisfied also, as may be seen by an analogous argument to that used above applied to sums of the form $\sum_{i=1}^{r} b_i e^{t_i x}$, which may be shown to have at most $(r - 1)$ zeros. $\qquad\square$

That there need not exist a best approximation to $f(x) \in C[a, b]$ from the class (9.5) is illustrated by the following example due to Meinardus (1967).

Example

$$[a, b] = [-1, 1], \, n = 2, \, f(x) = xe^x.$$

Let

$$\phi_\delta(x) = \frac{1}{\delta}e^{(1 + \delta)x} - \frac{1}{\delta}e^x.$$

Then

$$\| f(x) - \phi_\delta(x) \| \to 0 \qquad \text{as } \delta \to 0.$$

However, $f(x)$ does not belong to the class (9.5), and so no best approximation exists.

If best approximations are sought from the closure of the set (9.5), then the existence of a best approximation can be guaranteed, as shown by Rice (1962b). Unfortunately, Theorems 9.3 and 9.4 are then no longer valid. We will not pursue this further, but merely refer to the works of, for example, Braess (1967, 1970), Rice (1969), and Werner (1970).

Exercises

1. Show that the family of approximating functions defined by

$$\phi(x, \mathbf{a}) = \begin{cases} a_1 + a_2 e^x & a_2 \geq 1 \\ a_1 + (2 - a_2^{-x}) & a_2 < 1 \end{cases}$$

is unisolvent (Rice, 1964b).

2. Let

$$(f(x) - \phi(x, \mathbf{a}))(\phi(x, \mathbf{a} + \gamma \mathbf{c}) - \phi(x, \mathbf{a}) + o(\gamma)) > 0$$

for all $x \in \overline{E}(\mathbf{a})$. Prove that for $\gamma > 0$ sufficiently small,

$$\| f(x) - \phi(x, \mathbf{a} + \gamma \mathbf{c}) \| < \| f(x) - \phi(x, \mathbf{a}) \|.$$

3. Prove that the family (9.5) satisfies the representation condition.

4. Let

$$\phi(x, \mathbf{a}) = \frac{\displaystyle\sum_{i=1}^{m} a_i g_i(x)}{\displaystyle\sum_{i=m+1}^{n} a_i h_{i-m}(x)} = \frac{P(x, \mathbf{a})}{Q(x, \mathbf{a})}$$

where $Q(x, \mathbf{a}) > 0$ on $[a, b]$. Prove that this family of approximating functions satisfies the representation condition.

5. Prove that necessary conditions for $\phi(x, \mathbf{a})$ to be a best approximation to $f(x) \in C[a, b]$ are that there exist points $x_1, x_2, \ldots, x_t$ in $\overline{E}(\mathbf{a})$, where $t \leq n+1$, and a nontrivial vector $\lambda \in R^t$ such that

$$A^T \lambda = \mathbf{0}$$

$$\lambda_i \, \text{sign} \, (f(x_i) - \phi(x_i, \mathbf{a})) \geq 0 \qquad i = 1, 2, \ldots, t$$

where A is a $t \times n$ matrix with (i, j) element $\partial \phi(x_i, \mathbf{a})/\partial a_j$.

6. If the family $\phi(x, \mathbf{a})$ of approximating functions satisfies the representation condition, show that the conditions of exercise 5 are also sufficient for a best approximation.

9.3 Rational approximation on an interval

Let $f(x) \in C[a, b]$ and let $P(x)$ and $Q(x)$ be arbitrary polynomials of degree n and m respectively. Then the problem now considered is that of choosing the coefficients of the polynomials so that

$$\| f(x) - P(x)/Q(x) \| = \max_{a \leq x \leq b} |f(x) - P(x)/Q(x)|$$

is minimized. We may assume that $Q(x)$ has no zeros in $[a, b]$: any which exist must also be zeros of $P(x)$, and so these factors may be divided out. Assuming that any other common factors are also divided out, we arrive at a representation of the rational function $P(x)/Q(x)$ which is said to be *irreducible*: in other words, the only common factors are constants. The remaining degree of freedom may be used up by appropriate normalization of the coefficients; we will not, however, distinguish between two irreducible rational functions which may be made identical by suitable normalization. Without any loss of generality, then, we will assume the $Q(x) > 0$ on $[a, b]$. If ∂P denotes the (actual) degree of P (so that if $\partial P = k$, the coefficient of x^k in P is nonzero, but all coefficients of higher powers are zero) and ∂Q denotes the degree of Q, then best approximations are being sought from the set

$$R_{nm} = \{ P(x)/Q(x), \partial P \leqslant n, \partial Q \leqslant m,$$

$$P(x)/Q(x) \text{ is irreducible}, Q(x) > 0 \text{ on } [a, b]\}. \quad (9.7)$$

The irreducible representation of zero will be assumed to be $0/1$, and $\partial 0$ will be defined to be $-\infty$. The *defect* of $P(x)/Q(x) \in R_{nm}$ is defined to be $d = \min\{ n - \partial P, m - \partial Q \}$ and if $d > 0$ then the rational approximation is said to be *degenerate*, otherwise *nondegenerate* or *normal*. Degeneracy corresponds to the (nominally) degree n and m polynomials $P(x)$ and $Q(x)$ having at least one equal factor which is not just a constant. For example, if $n = 2, m = 3, a = 0, b = 1$, we may think of $1/(1 + x)$ as an element of R_{23} having defect 2.

We begin the study of approximations from the class (9.7) by considering the problem of existence of a best approximation. This, of course, is not dealt with by the analysis of Chapter 1; however, that an existence result is indeed available is demonstrated by the following theorem, due to Walsh (1931).

Theorem 9.6 If $f(x) \in C[a, b]$, then there exists a best approximation to $f(x)$ from R_{nm}.

Proof Let

$$\delta = \inf_{R \in R_{nm}} \| f - R \|,$$

and let $R_k \in R_{nm}$ be a sequence of approximations such that

$$\| f - R_k \| \to \delta.$$

We can write $R_k = P_k/Q_k$, normalized so that $\| Q_k \| = 1$. Thus

$$\| R_k \| \leqslant \| R_k - f \| + \| f \|$$

$$\leqslant \delta + 1 + \| f \| = M, \text{ say,}$$

for k sufficiently large, say $k \geqslant k_0$. Now

$$\| P_k \| \leqslant \| R_k Q_k \| \leqslant M \qquad k \geqslant k_0,$$

and so by Lemma 1.1 (considering subsequences if necessary, which we do not rename)

$$P_k \to P$$
$$Q_k \to Q$$

where $\|P\| \leqslant M$ and $\|Q\| = 1$. There are at most m points where $Q(x) = 0$ and at all other points P/Q is well-defined and we have

$$\frac{P_k}{Q_k} \to \frac{P}{Q}.$$

Further, $|P_k/Q_k| \leqslant M$ in $[a, b]$, $k \geqslant k_0$, so for the points where $Q \neq 0$ it follows that $|P/Q| \leqslant M$ on taking limits. By continuity the last inequality must hold for all $x \in [a, b]$, and so any zero of Q in $[a, b]$ must also be a zero of P, and a corresponding linear factor may be cancelled from P and Q. Eventually, therefore, Q will be free of zeros in $[a, b]$ and P/Q can be regarded now as an element, say R, of R_{nm}. Thus

$$\|f - R\| \leqslant \|f - R_k\| + \|R_k - R\|$$
$$= \delta$$

on taking limits as $k \to \infty$. But $\|f - R\| \geqslant \delta$ by definition of δ and so $\|f - R\| = \delta$ and R is a best approximation. □

The presence of the linear terms in the numerator and denominator of the elements of R_{nm} plays a vital role in the existence proof. However, linearity alone is not enough, for the cancellation of factors makes specific use of the polynomial nature of P and Q. More general rational expressions formed by quotients of elements from arbitrary finite dimensional linear subspaces of $C[a, b]$ cannot be treated in this manner, and in general existence of best approximations by such generalized rationals cannot be guaranteed (although see Boehm, 1965).

Example Consider the approximation of $f(x) = x$ by functions of the form $R = ax^2/(b + cx)$ on $[0, 1]$ with $b + cx > 0$ on $[0, 1]$. Setting $a = 1, c = 1$, we have $\|f - R\| \to 0$ as $b \to 0$ but $f(x)$ can not be represented by an approximating function of the required form.

From the point of view of characterization, however, the linearity of the numerator and denominator is the important property which is to be exploited, and so it is convenient in what follows to widen the class of approximating functions. In particular, let us define

$$\overline{R}_{nm} = \{ P(x)/Q(x), P(x) = \sum_{i=0}^{n} a_i g_i(x), Q(x) = \sum_{i=0}^{m} b_i h_i(x),$$
$$g_i(x) \in C[a, b], i = 0, 1, \ldots, n, h_i(x) \in C[a, b],$$
$$i = 0, 1, \ldots m, Q(x) > 0 \text{ on } [a, b]\}.$$

For given $R \in \overline{R}_{nm}$, let S_{nm} denote the linear subspace of $C[a, b]$ spanned by

$$\{g_0(x), g_1(x), \ldots, g_n(x), Rh_0(x), \ldots, Rh_m(x)\}$$

and let the dimension of S_{nm} be k; clearly we must have $k \leqslant n + m + 1$ (exercise 7). Further, define $\overline{E}$ by

$$\overline{E} = \{x: x \in [a, b], |f(x) - R(x)| = \|f - R\|\}$$

and let $\theta(x) = \text{sign}\,(f(x) - R(x))$, $x \in \overline{E}$. The following theorem generalizes (and so contains as a special case) Theorem 3.1.

Theorem 9.7   Let $f(x) \in C[a, b]$. Then $R \in \overline{R}_{nm}$ is a best approximation to $f(x)$ if and only if there exists $E \subset \overline{E}$ consisting of $t \leqslant k + 1$ points $x_1, x_2, \ldots, x_t$ and a nontrivial vector $\lambda \in R^t$ such that

$$\sum_{i=1}^{t} \lambda_i S(x_i) = 0 \qquad \text{for all } S \in S_{nm}$$

$$\lambda_i \theta_i \geqslant 0 \qquad i = 1, 2, \ldots, t,$$

where $\theta_i = \theta(x_i)$, $i = 1, 2, \ldots, t$.

Proof   (If $\|f - R\| = 0$ the result is trivial, so assume not.)
Sufficiency Let the conditions be satisfied at $R \in \overline{R}_{nm}$, and let $R_1 = P_1/Q_1 \in \overline{R}_{nm}$ give a better approximation. Then for all $x \in \overline{E}$,

$$|f(x) - R_1(x)| < |f(x) - R(x)|$$

so that in particular

$$\theta_i(R_1(x_i) - Rx_i)) > 0 \qquad i = 1, 2, \ldots, t$$

or

$$\theta_i(P_1(x_i) - R(x_i)Q_1(x_i)) > 0 \qquad i = 1, 2, \ldots, t.$$

Now $S(x) = P_1(x) - R(x)Q_1(x) \in S_{nm}$. Thus the conditions of the theorem lead to a contradiction, and the result follows.
Necessity 1 The conditions of Theorem 1.6 give

$$\langle g_j/Q, v \rangle = 0 \qquad j = 0, 1, \ldots, n,$$
$$\langle h_j R/Q, v \rangle = 0 \qquad j = 0, 1, \ldots, m,$$

where

$$\langle u, v \rangle = \int_a^b uv \, dx$$

and where

$$v \in \text{conv}\,\{\theta(x)\delta(x), x \in \overline{E}\}.$$

Thus there exist $t \leqslant k+1$ points $x_1, x_2, \ldots, x_t$ in $\overline{E}$ such that

$$\sum_{i=1}^{t} \mu_i \frac{S(x_i)\theta_i}{Q(x_i)} = 0$$

where

$$\sum_{i=1}^{t} \mu_i = 1 \qquad \mu_i \geqslant 0, \, i = 1, 2, \ldots, t.$$

Setting

$$\lambda_i = \frac{\mu_i \theta_i}{Q(x_i)} \qquad i = 1, 2, \ldots, t,$$

the result follows.

Necessity 2 Let R be a best approximation, but the conditions of the theorem not be satisfied. Then

$$0 \notin \text{conv } \{\theta(x)S(x), \, x \in \overline{E}, \, S(x) \in S_{nm}\}.$$

By Theorem 1.5 (with $n = k$), there exists $S_1(x) \in S_{nm}$ such that

$$\theta(x)S_1(x) > 0 \qquad \text{for all } x \in \overline{E}.$$

Thus there exists $P_1(x)$, $Q_1(x)$ such that

$$\theta(x)(P_1(x) + R(x)Q_1(x)) > 0 \qquad \text{for all } x \in \overline{E}$$

or

$$\theta(x)(P_1(x)Q(x) + P(x)Q_1(x)) > 0 \qquad \text{for all } x \in \overline{E}.$$

Now define

$$R_\gamma = \frac{P + \gamma P_1}{Q - \gamma Q_1} \qquad \gamma > 0$$

so that $R_\gamma \in \overline{R}_{nm}$ for $\gamma > 0$ sufficiently small. Then

$$f - R_\gamma = f - R + \frac{P}{Q} - \frac{P + \gamma P_1}{Q - \gamma Q_1}$$

$$= f - R - \gamma \frac{P_1 Q + P Q_1}{Q(Q - \gamma Q_1)},$$

where the last term on the right hand side has the sign of $f - R$ at all points of $\overline{E}$. Thus, using exercise 3 of Chapter 3, for $\gamma > 0$ sufficiently small

$$\|f - R_\gamma\| < \|f - R\|$$

which gives a contradiction, and concludes the proof. $\qquad \square$

Corollary Let $R \in \overline{R}_{nm}$ be a best approximation to $f(x) \notin \overline{R}_{nm}$, and let S_{nm} be a

Haar subspace of dimension k. Then if for some $S(x) \in S_{nm}$,

$$\theta(x)S(x) \geq 0 \qquad \text{for all } x \in \overline{E},$$

$S(x)$ must be identically zero.

Proof In the theorem we must have $t = k+1$, and

$$\mu_i = \lambda_i \theta_i > 0, \ i = 1, 2, \ldots, k+1.$$

Therefore at least one of the numbers $\theta_i S(x_i)$ is positive and one is negative if $S(x)$ is not identically zero. □

Theorem 9.8 Let $R \in \overline{R}_{nm}$ be a best approximation to $f(x)$, and let S_{nm} be a Haar subspace of dimension k. Then R is unique.

Proof Let $R_1 = P_1/Q_1 \in \overline{R}_{nm}$ be another best approximation. Then $f \notin \overline{R}_{nm}$. Also since

$$\theta(x)((f(x) - R(x)) - (f(x) - R_1(x))) \geq 0 \qquad \text{for all } x \in \overline{E}$$

it follows that

$$\theta(x)(R_1(x) - R(x)) \geq 0 \qquad x \in \overline{E}$$

and so

$$\theta(x)(P_1(x) - Q_1(x)R(x)) \geq 0 \qquad x \in \overline{E}.$$

The result follows from the corollary to Theorem 9.7. □

Returning now to the case of ordinary polynomial rational approximation, the conditions of Theorem 9.7 can be given in a particularly convenient and useful form which has strong similarities (and indeed contains as a special case) the corresponding result for polynomial approximation. This theorem was first given in this form by Achieser in 1930 (in Russian).

Theorem 9.9 Let $f(x) \in C[a, b]$. Then $R \in R_{nm}$ is a best approximation to $f(x)$ on $[a, b]$ if and only if there exists an alternating set of $N = n + m + 2 - d$ points in $[a, b]$, where d is the defect of R.
  (If $\| f - R \| = 0$ the result is trivial, so assume not.)

Proof 1 The subspace S_{nm} is spanned by

$$\{1, x, x^2, \ldots, x^n, R(x), R(x)x, \ldots, R(x)x^m\}.$$

Thus S_{nm} is a Haar subspace of $C[a, b]$ of dimension $m + n + 1 - d$, and a basis for S_{nm} is given by

$$\{x^i/Q(x), i = 0, 1, \ldots, m + n - d\}$$

(see exercise 11). In Theorem 9.7, we must have $t = N$, with $\lambda_i \neq 0, i = 1, 2, \ldots,$ N, and so the result follows using Lemma 3.1.

Proof 2 Suppose first that $R \in R_{nm}$, and that such an alternating set exists, so that

$$|f(x_i) - R(x_i)| = \| f - R \| \qquad i = 1, 2, \ldots, N.$$

If $R_1 \in R_{nm}$ is a better approximation, then

$$R_1 - R = (f - R) - (f - R_1)$$

alternates in sign as we go through $x_1, x_2, \ldots, x_N$ and so has $N - 1$ zeros in $[a, b]$. But the numerator of $R_1 - R$ is a polynomial of degree $\leqslant m + n - d = N - 2$. This gives a contradiction, and so the sufficiency of the conditions is proved.

Now let $R \in R_{nm}$ be a best approximation, but the conditions of the theorem not be satisfied, so that the largest alternating set consists of $N_1 < N$ points. As in the polynomial (Chebyshev set) case, we define $N_1 - 1$ points $z_1 < z_2 < \ldots < z_{N_1 - 1}$ in $[a, b]$ where $f = R$, and a polynomial

$$\overline{P}(x) = \pm \prod_{i=1}^{N_1 - 1} (x - z_i)$$

whose sign is that of $f - R$ in all sufficiently small intervals I_j of $[a, b]$ containing all the points where $\| f - R \|$ is attained. Since P and Q have no factors other than constants, there exist polynomials P_1 and Q_1 satisfying

$$\overline{P} = QP_1 - Q_1 P,$$

(exercise 9). For $\gamma > 0$ sufficiently small,

$$R_\gamma = \frac{P + \gamma P_1}{Q + \gamma Q_1} \in R_{nm},$$

and further

$$f - R_\gamma = f - R + \frac{P}{Q} - \frac{P + \gamma P_1}{Q + \gamma Q_1}$$

$$= f - R - \gamma \frac{\overline{P}}{Q(Q + \gamma Q_1)},$$

with the last term on the right hand side having the same sign as $f - R$ in the intervals I_j introduced above. An argument similar to that used in Theorem 3.2 gives a contradiction to the assumption that R is a best approximation and concludes the proof. $\qquad \square$

If R is a best approximation which is normal, then the alternating set must contain $n + m + 2$ points. There are, however, many examples of degenerate best approximations.

Example

$$f(x) = x^2, \qquad [a, b] = [a, 1] \qquad m = n = 1$$

For values of a satisfying $-1 < a < 1$, the (unique) best approximation R is normal, with 4 points in the alternating set. For instance, when $a = -0.8$, we have (to six decimal places)

$$R = \frac{0.324\ 762 - 0.296\ 299x}{1 - 0.978\ 517x}$$

with $\|f - R\| = 0.324\ 880$, and alternating set -0.8, $0.010\ 977$, $0.910\ 977$, 1.0. When $a = -1$, we have $\|x^2 - \frac{1}{2}\| = \frac{1}{2}$ with

$$f(-1) - \tfrac{1}{2} = -(f(0) - \tfrac{1}{2}) = f(1) - \tfrac{1}{2} = \tfrac{1}{2},$$

so that $f - \frac{1}{2}$ has an alternating set of 3 points. Further $R = \frac{1}{2}$ has defect 1, and so is a best approximation.

This example suggests that it would be of interest to see what happens as $a \to -1$. For $a = -0.999$, the best approximation is

$$R = \frac{0.499\ 001 - 0.499\ 000x}{1 - 0.999\ 999x}$$

with $\|f - R\| = 0.499\ 001$ and alternating set

$$-0.999\ 000,\ 0.000\ 000,\ 0.999\ 500,\ 1.0.$$

In fact (see Figure 18) two points of the alternating set are coalescing as $a \to -1$, although the errors remain of opposite sign. The structure of the error curve suggests that we should expect difficulties in computing degenerate or nearly degenerate best rational approximations; we return to this point later.

The close adherence to the polynomial theory which is illustrated by Theorem 9.9 is maintained in the question of uniqueness. The following result (also due to Achieser) is an immediate consequence of Theorem 9.8; we give an alternative direct proof.

Theorem 9.10 The best approximation to $f(x) \in C[a, b]$ from R_{nm} is unique.

Proof Let R and R_1 be best approximations. Then by Theorem 9.9, $f - R$ has an alternating set of points $x_1, x_2, \ldots, x_N$ and $f - R_1$ has an alternating set of points $y_1, y_2, \ldots, y_{N_1}$. Assume that $N \geqslant N_1$, and let

$$G = R_1 - R = (f - R) - (f - R_1).$$

Now if $G(x_i) \neq 0$, sign $(G(x_i)) = $ sign $(f(x_i) - R(x_i))$ and if $G(x_i) \neq 0$, $G(x_{i+j}) \neq 0$, then sign $(G(x_i)) = (-1)^j$ sign $(G(x_{i+j}))$ by the alternation property. Further, the zeros of G are zeros of its numerator $P_1 Q - P Q_1$ which is a

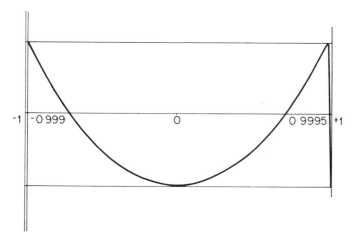

-1 -0·999 0 0·9995 +1

Figure 18

polynomial of degree at most $m + n - d_1 = N_1 - 2 \leqslant N - 2$. This can have at most $N - 2$ zeros in $[a, b]$; we will show that this fact is contradicted.

If j is odd, $G(x)$ has an odd number of zeros in $[x_i, x_{i+j}]$ (counting multiple zeros according to multiplicity). If j is even, $G(x)$ has an even number of zeros in $[x_i, x_{i+j}]$. However, $G(x)$ has at least $(j - 1)$ zeros in $[x_i, x_{i+j}]$ by the sign alternation property, and so $G(x)$ must have at least j zeros in $[x_i, x_{i+j}]$. It follows that $G(x)$ must have at least $N - 1$ zeros in $[a, b]$, which gives the required contradiction, and shows that $R = R_1$. □

The best approximation R^* from R_{nm} is in fact strongly unique, in the sense that there exists $\gamma > 0$ such that

$$\|f - R\| \geqslant \|f - R^*\| + \gamma\|R - R^*\|$$

for all $R \in R_{nm}$. This result was first given by Cheney and Loeb (1964); the following proof is from Cheney (1966).

Theorem 9.11 The best approximation to $f(x) \in C[a, b]$ from R_{nm} is strongly unique.

Proof If $f \in R_{nm}$, we can take $\gamma = 1$, so assume not. Define

$$\gamma(R) = \frac{\|f - R\| - \|f - R^*\|}{\|R - R^*\|},$$

where R^* is the best approximation, and assume that there exists a sequence $\{R_k\} \in R_{nm}$ such that $R_k \neq R^*$ and $\gamma(R_k) \to 0$ as $k \to \infty$. Let the approximations $R_k = P_k/Q_k$ be normalized so that $\|Q_k\| = 1$, and let $\|Q^*\| = 1$ where $R^* = P^*/Q^*$. Now since $\gamma(R_k) \to 0$, $\|R_k\|$ is bounded so that, by compactness, there exist subsequences (which we do not rename) such that

$$P_k \to P$$
$$Q_k \to Q$$

as $k \to \infty$. Let $x \in \overline{E}$. Then if $\theta(x) = \operatorname{sign}(f(x) - R^*(x))$, $x \in \overline{E}$,

$$
\begin{aligned}
\gamma(R_k)\|R^* - R_k\| &= \|f - R_k\| - \|f - R^*\| \\
&\geqslant \theta(x)(f(x) - R_k(x)) - \theta(x)(f(x) - R^*(x)) \\
&= \theta(x)(R^*(x) - R_k(x)) \\
&= -\frac{\theta(x)}{Q_k(x)}(P_k(x) - Q_k(x)R^*(x)).
\end{aligned}
\tag{9.8}
$$

Letting $k \to \infty$ we must have

$$\theta(x)S(x) \geqslant 0 \qquad \text{for all } x \in \overline{E}$$

where $S = P - QR^* \in S_{nm}$ and so by the corollary to Theorem 9.7, $S \equiv 0$, showing that $R^* = P/Q$ so that $P^* = P$, $Q^* = Q$. Now let

$$\delta = \min_{\substack{S \in S_{nm} \\ \|S\| = 1}} \max_{x \in \overline{E}} \theta(x)S(x)$$

which is positive by the corollary to Theorem 9.7. Thus there exists $x \in \overline{E}$ such that from (9.8)

$$
\begin{aligned}
\gamma(R_k)\|R^* - R_k\| &\geqslant \frac{\theta(x)}{Q_k(x)}(Q_k(x)R^*(x) - P_k(x)) \\
&\geqslant \theta(x)(Q_k(x)R^*(x) - P_k(x))
\end{aligned}
$$

since $0 < Q_k(x) \leqslant 1$,

$$
\begin{aligned}
&\geqslant \delta\|Q_k R^* - P_k\| \\
&\geqslant \delta\varepsilon\|R^* - R_k\|
\end{aligned}
$$

for some $\varepsilon > 0$ and k sufficiently large, since $Q_k \to Q^*$ implies that we must have Q_k eventually bounded away from zero on $[a, b]$. Since $R_k \neq R^*$, the last inequality gives a contradiction on letting $k \to \infty$. Thus $\gamma(R)$ must be bounded away from zero, and the result is proved. $\qquad\square$

Finally, in this section, we draw attention to the fact that theorems of Jackson and Muntz–Jackson type exist for rational approximation. Let $0 \leqslant \lambda_1 < \lambda_2 < \ldots < \lambda_n$ be real numbers, and let R_Λ denote rational functions of the form

$$R(x) = \frac{\displaystyle\sum_{i=1}^{n} a_i x^{\lambda_i}}{\displaystyle\sum_{i=1}^{n} b_i x^{\lambda_i}}.$$

Bak and Newman (1978) have shown that for any infinite sequence $\lambda_1, \lambda_2, \ldots$ the set of rational functions $R(x) \in R_\Lambda$ is dense in $C[0, 1]$ provided only that the sequence $\{\lambda_k\}$ is bounded away from zero. This is an improvement of an earlier result which required that $\lambda_n \to \infty$. Recent error estimates for approximation by elements of R_Λ to general functions $f(x)$ are given by Bak (1977). Many results have also been obtained for particular functions $f(x)$. For example, the case $f(x) = |x|$ treated by Bernstein (1912) in the polynomial case has been considered by Newman (1964). When $\lambda_i = i, i = 1, 2, \ldots, n$, and the interval is $[-1, 1]$, he has shown that

$$\| |x| - R(x) \| < 3e^{-n^{\frac{1}{2}}} \qquad n \geqslant 4.$$

Other estimates have been obtained for $f(x) = e^x, e^{-x}$. A good survey of these results (and many others) is given by Reddy (1978).

Exercises

7. Show that the dimension of the linear space S_{nm} is at most $n + m + 1$.

8. By considering rational functions of the form $1/(ax + 1)$ which belong to $C[0, 1]$, demonstrate the loss of compactness.

9. Let Π_n denote the space of polynomials of degree n, and let $P(x) \in \Pi_n$, $Q(x) \in \Pi_m$ be given, where (without loss of generality) $\partial P = n$, $\partial Q = m$.

(i) Prove that for $P(x)$ and $Q(x)$ to have a (nonconstant) factor, it is necessary and sufficient that two polynomials $g(x)$ and $h(x)$ (not both zero) can be found such that

$$h(x)Q(x) = g(x)P(x)$$

where $\partial g(x) \leqslant m - 1$, $\partial h(x) \leqslant n - 1$. Deduce that an equivalent necessary and sufficient condition is that the determinant R of a certain $(m + n) \times (m + n)$ matrix be zero.

(ii) Prove that if $P(x)$ and $Q(x)$ are irreducible, there exist polynomials $P_1(x) \in \Pi_n$, $Q_1(x) \in \Pi_m$ such that

$$P_1(x)Q(x) - P(x)Q_1(x) = \overline{P}(x)$$

where $\overline{P}(x) \in \Pi_{m+n}$. (Use part (i) to show that a certain $(m + n + 1) \times (m + n + 2)$ matrix has full rank.)

10. Rigorously complete the second necessity proof of Theorem 9.9.

11. Let $R \in R_{nm}$ have defect d. Prove that in this case S_{nm} is a Haar subspace of $C[a, b]$ of dimension $n + m + 1 - d$, with basis $\{x^j/Q(x), j = 0, 1, \ldots, n + m - d\}$ where $R = P/Q$.

12. Let $R \in \overline{R}_{nm}$, and let $\{g_0(x), \ldots, g_n(x)\}$ and $\{h_0(x), \ldots, h_m(x)\}$ form Chebyshev sets on $[a, b]$. Show that S_{nm} need not be a Haar subspace of $C[a, b]$ by considering the example on $[0, 3]$ with $n = m = 1$, $g_0(x) = 1$, $g_1(x) = x^2$, $h_0(x)$

= 1, $h_1(x) = x$, $R = 11 + x^2$, and finding an element $S(x) \in S_{11}$ with zeros at $x = 1$, 2, and 3.

13. Determine the best Chebyshev approximation to $f(x) = x + \alpha$ on $[-1, 1]$ of the form $1/(a + bx)$. Sketch the error curve, and explain what happens as $\alpha \to 0$.

14. Let $R \in R_{nm}$ be the best Chebyshev approximation to $f(x)$ on $[-1, 1]$. If $f(x)$ is even (odd) show that R must also be even (odd). Does this result hold for any of the other L_p norms on $C[a, b]$?

15. Use exercise 14 to obtain the best Chebyshev approximations from R_{11} on $[-1, 1]$ to (i) $f(x) = x^3$, (ii) $f(x) = x^4$.

16. Let $S_n(x)$ be a spline function of degree n with knots $x_1 < x_2 < \ldots < x_N$. Prove that $S_n(x)/S_m(x)$ is a piecewise rational function, and find the amount of continuity at the knots. Determine the number of degrees of freedom of a *general* piecewise rational function with N knots, which belongs to $C^l[a, b]$, where $a < x_1$ and $x_N < b$. Find the unique piecewise rational R formed by two rational functions from R_{11}, with one knot at $x = 0$, such that $R \in C^1[-1, 1]$ and R satisfies the interpolation conditions

$$R(-1) = 2, \ R(0) = 1, \ R(1) = 2, \ DR(0) = 4.$$

17. Let R be an approximation from R_{nm} to $f(x)$, with defect d, and let there exist $k = n + m + 2 - d$ points in $[a, b]$ with

$$f(x_i) - R(x_i) = \rho_i \neq 0 \qquad i = 1, 2, \ldots, k$$
$$\text{sign}(\rho_{i+1}) = -\text{sign}(\rho_i) \qquad i = 1, 2, \ldots, k-1.$$

Prove that

$$\min_i |\rho_i| \leqslant \inf_{R \in R_{nm}} \|f - R\|.$$

(cf. exercise 15 of Chapter 3).

18. Let $X = [-1, 1] \times [-1, 1]$ and consider the problem of approximating by rational functions of the two variables x and y of the form

$$R = \frac{P}{Q} = \frac{\sum_{i=0}^{n} \sum_{k=0}^{v_i} a_{ik} x^i y^k}{\sum_{j=0}^{m} \sum_{l=0}^{u_i} b_{jl} x^j y^l} \qquad Q > 0 \text{ on } X.$$

Demonstrate non-existence of the best approximation for the case $n = 2$, $m = 1$, $v_i = 2$, $i = 0, 1, 2$, $u_j = 1$, $j = 0, 1$, to the function

$$f(x) = \begin{cases} \dfrac{(x+1)^2 + (y+1)^2}{x + y + 2} & -1 \leqslant x \leqslant 1, \ -1 < y \leqslant 1 \\ x + 1 & -1 \leqslant x \leqslant 1, \ y = -1. \end{cases}$$

(Henry and Weinstein, 1974).

19. Demonstrate non-existence of the best rational L_p approximation, $1 \leqslant p < \infty$, by considering the approximation of $f(x) = x^{-\frac{1}{2}p}$ by functions of the form

$$\frac{a}{b + cx^{\frac{1}{2}p}} \qquad b + cx^{\frac{1}{2}p} > 0, \ x \in [0, 1]$$

in the interval $[0, 1]$ (Wolfe, 1974).

9.4 Rational approximation on a discrete set

We turn now to the discrete problem corresponding to the (polynomial) rational continuous problem dealt with in the previous section. Let $X = \{x_1, x_2, \ldots, x_t\} \in [a, b]$ and define

$$R_{nm}{}^D = \{P(x)/Q(x), \ \partial P \leqslant n, \ \partial Q \leqslant m, \ P(x)/Q(x) \text{ is irreducible}, \\ Q(x_i) > 0, \ i = 1, 2, \ldots, t\}. \qquad (9.9)$$

Then we seek $R \in R_{nm}{}^D$ to minimize

$$\|f - R\| = \max_{1 \leqslant i \leqslant t} |f(x_i) - R(x_i)| \qquad (9.10)$$

for some numbers $f(x_1), \ldots, f(x_t)$. Notice that here the norm is one on the finite dimensional space R^t, and not on $C[a, b]$; f and R appear on the left hand side of (9.10) purely for notational convenience, and $f - R$ in fact here represents a vector in R^t with components $f(x_i) - R(x_i)$, $i = 1, 2, \ldots, t$. In addition, there need be no actual function $f(x)$ defined, merely a vector in R^t, for example of data values.

There are certain fundamental differences between this problem and the previous one: for example the existence of a best approximation is no longer guaranteed.

Example $f(x) = 1 - \frac{3}{2}x + \frac{1}{2}x^2$ $\qquad X = \{0, 1, 2\}$ $\qquad n = 0, m = 1$

$$f(x_1) - \frac{a}{1 + bx_1} = 1 - a$$

$$f(x_2) - \frac{a}{1 + bx_2} = -\frac{a}{1 + b}$$

$$f(x_3) - \frac{a}{1 + bx_3} = -\frac{a}{1 + 2b}$$

Choosing $a = 1$, we have

$$\left\| f - \frac{a}{1 + bx} \right\| \to 0 \text{ as } b \to \infty.$$

Thus the best approximation must satisfy

$$\frac{1}{1+b} = \frac{1}{1+2b} = 0$$

which is impossible. Thus no best approximation exists.

The analogue of Theorem 9.9 is also no longer valid; the proof of necessity still goes through, but the conditions are no longer sufficient because the sign changes of $R_1 - R$ may be at points where the numerator is nonzero, if the denominator has a zero in $[a, b] - X$. (If $[a, b]$ is not given explicitly, we may consider it to be the smallest interval containing X.)

Example $X = \{-1, -\frac{1}{2}, 1\}, f(x_1) = -1, f(x_2) = -2, f(x_3) = 1, n = 0, m = 1.$
The approximation

$$R = \frac{-9}{16x + 20} \in R_{nm}{}^D$$

and is such that

$$[f(-1) - R(-1)] = -[f(-\tfrac{1}{2}) - R(-\tfrac{1}{2})] = [f(1) - R(1)] = 5/4,$$

so that there exists an alternating set of $m + n + 2 - d = 3$ points. However, $1/x \in R_{nm}{}^D$ is clearly a better approximation than R: the difficulty here is that $1/x \notin R_{nm}$.

The possibility of zeros in the denominator of $R_1 - R$ in $[a, b] - X$ also invalidates the proof of uniqueness. It is clear that in order to obtain analogues of Theorems 9.9 and 9.10, it is necessary to seek approximations out of R_{nm}, and not $R_{nm}{}^D$.

We turn now to the provision of algorithms for computing best approximations from R_{nm} and $R_{nm}{}^D$. Two important basic methods are considered, both of which may be applied to either the discrete or continuous problem. The first of these, however, lends itself more naturally to the discrete problem, as then difficult subproblems are avoided, and we describe the method as it applies to approximation from $R_{nm}{}^D$.

Exercises

20. A *seminorm* on a linear space is a function of the elements which satisfies the conditions (ii) and (iii) of Definition 1.1. Show that if $f(x) \in C[a, b]$, the right hand side of equation (9.10) is a seminorm on $C[a, b]$.

21. Let $X \cdot = \{-1, 0, 1\}, n = 0, m = 1$, and $f(x) = x$. By satisfying the necessary conditions for a best approximation, find the minimum value of the norm, and show that there are two best approximations.

22. By satisfying the necessary conditions, determine the two approximations for the example given above.

9.5 The differential correction algorithm

The differential correction algorithm was first described by Cheney and Loeb in 1961. A modified version of the method was subsequently considered by Cheney and Loeb (1962) and Cheney and Southard (1963), and this was shown to have sure convergence properties, thus directing attention away from the original method. Recently, however, Barrodale, Powell, and Roberts (1972) have studied both methods for the case of approximation on discrete sets, and have shown that the method in its original form has not only guaranteed convergence from any starting point in $R_{nm}{}^D$, but is frequently quadratically convergent. In a comparison of the original and modified versions they illustrate that the former is to be preferred.

Let $f(x_1), f(x_2), \ldots, f(x_t)$ be given, and let $\Delta^* = \inf_{R \in R_{nm}{}^D} \| f - R \|$. Then the original method can be defined as follows

(1) Choose as initial approximation $R_1 = P_1/Q_1 \in R_{nm}{}^D$; set $k = 1$.
(2) Determine polynomials $P(x)$ and $Q(x)$ of degree n and m respectively to minimize

$$\max_{1 \leqslant i \leqslant t} \left\{ \frac{|f(x_i)Q(x_i) - P(x_i)| - \Delta_k Q(x_i)}{Q_k(x_i)} \right\} \qquad (9.11)$$

where

$$\Delta_k = \max_{1 \leqslant i \leqslant t} |f(x_i) - P_k(x_i)/Q_k(x_i)| \qquad (9.12)$$

and $Q(x)$ is not identically zero.
(3) Unless $\Delta_k = \Delta^*$, set $P_{k+1} = P$, $Q_{k+1} = Q$, $k = k+1$ and go to (1).

The motivation for the method is as follows. For any choice of $P(x)$ and $Q(x)$ with $Q(x_i) > 0$, $i = 1, 2, \ldots, t$, let Δ be such that

$$|f(x_i) - P(x_i)/Q(x_i)| \leqslant \Delta \qquad i = 1, 2, \ldots, t. \qquad (9.13)$$

Then

$$|f(x_i)Q(x_i) - P(x_i)| \leqslant \Delta Q(x_i) \qquad i = 1, 2, \ldots, t. \qquad (9.14)$$

Now for any function of two variables $g(\alpha, \beta)$ which is sufficiently differentiable, we can write the Taylor series expansion about α_k and β_k as

$$g(\alpha, \beta) = g(\alpha_k, \beta_k) + (\alpha - \alpha_k)\frac{\partial g}{\partial \alpha_k} + (\beta - \beta_k)\frac{\partial g}{\partial \beta_k} + \ldots$$

Thus, letting $g(\Delta, Q(x_i)) = \Delta Q(x_i)$, for given i, it follows that

$$\Delta Q(x_i) = \Delta_k Q_k(x_i) + (\Delta - \Delta_k)Q_k(x_i) + (Q(x_i) - Q_k(x_i))\Delta_k + \cdots$$

so to first order terms, (9.14) can be written

$$\begin{aligned}
|f(x_i)Q(x_i) - P(x_i)| &\leqslant \Delta_k Q_k(x_i) + (\Delta - \Delta_k)Q_k(x_i) \\
&\quad + (Q(x_i) - Q_k(x_i))\Delta_k && i = 1, 2, \ldots, t \\
&= (\Delta - \Delta_k)Q_k(x_i) + Q(x_i)\Delta_k && i = 1, 2, \ldots, t.
\end{aligned}$$

Thus, to first order terms,

$$|f(x_i)Q(x_i) - P(x_i)| - \Delta_k Q(x_i) \leqslant (\Delta - \Delta_k)Q_k(x_i) \qquad i = 1, 2, \ldots, t$$

or

$$\max_{1 \leqslant i \leqslant t} \left\{ \frac{|f(x_i)Q(x_i) - P(x_i)| - \Delta_k Q(x_i)}{Q_k(x_i)} \right\} + \Delta_k \leqslant \Delta.$$

It follows that Δ is *approximately* minimized by finding the minimum of the expression on the left hand side of the above inequality, in other words the minimum of (9.11).

We now examine the convergence properties of the algorithm, first showing that the sequence Δ_k is monotonically decreasing, and also that successive approximations $R_k \in R_{nm}{}^D$. It will be assumed that $Q_k(x)$ is normalized at each step by making the maximum modulus coefficient unity.

Theorem 9.12 If $Q_k(x_i) > 0$, $i = 1, 2, \ldots, t$ and if $\Delta_k \neq \Delta^*$, then $\Delta_{k+1} < \Delta_k$ and $Q_{k+1}(x_i) > 0$, $i = 1, 2, \ldots, t$.

Proof Since $\Delta_k \neq \Delta^*$ there exists $\overline{R} = \overline{P}/\overline{Q} \in R_{nm}{}^D$ with $\overline{Q}$ normalized so that the maximum modulus coefficient is 1, such that

$$\|f - \overline{R}\| = \overline{\Delta} < \Delta_k.$$

Let $\delta = \min\limits_{1 \leqslant i \leqslant t} \overline{Q}(x_i)$, and let $M = \max\limits_{1 \leqslant i \leqslant t} \sum\limits_{j=0}^{n} |x_i|^j$, so that $Q(x_i) \leqslant M$, $i = 1$, $2, \ldots, t$ for any polynomial $Q(x)$ of degree m satisfying $Q(x_i) > 0, i = 1, 2, \ldots, t$, and the assumed normalization that the maximum modulus coefficient is 1. Now, by the definition of P_{k+1} and Q_{k+1},

$$\max_{1 \leqslant i \leqslant t} \left\{ \frac{|f(x_i)Q_{k+1}(x_i) - P_{k+1}(x_i)| - \Delta_k Q_{k+1}(x_i)}{Q_k(x_i)} \right\}$$

$$\leqslant \max_{1 \leqslant i \leqslant t} \left\{ \frac{|f(x_i)\overline{Q}(x_i) - \overline{P}(x_i)| - \Delta_k \overline{Q}(x_i)}{Q_k(x_i)} \right\}$$

$$\leqslant \max_{1 \leqslant i \leqslant t} \left\{ \left[\left| f(x_i) - \frac{\overline{P}(x_i)}{\overline{Q}(x_i)} \right| - \Delta_k \right] \frac{\overline{Q}(x_i)}{Q_k(x_i)} \right\}$$

$$\leq \max_{1 \leq i \leq t} \left\{ [\overline{\Delta} - \Delta_k] \frac{\overline{Q}(x_i)}{Q_k(x_i)} \right\}$$

$$= -[\Delta_k - \overline{\Delta}] \min_{1 \leq i \leq t} \frac{\overline{Q}(x_i)}{Q_k(x_i)}$$

$$\leq -\frac{\delta}{M}[\Delta_k - \overline{\Delta}]. \tag{9.15}$$

Thus we must have

$$\Delta_k Q_{k+1}(x_i)/Q_k(x_i) \geq \delta[\Delta_k - \overline{\Delta}]/M > 0 \qquad i = 1, 2, \ldots, t \tag{9.16}$$

showing that

$$Q_{k+1}(x_i) > 0 \qquad i = 1, 2, \ldots, t.$$

Further, the above string of inequalities leading to (9.15) shows that

$$\{ |f(x_i)Q_{k+1}(x_i) - P_{k+1}(x_i)| - \Delta_k Q_{k+1}(x_i) \}/Q_k(x_i) < 0 \qquad i = 1, 2, \ldots, t$$

and so

$$|f(x_i) - P_{k+1}(x_i)/Q_{k+1}(x_i)| < \Delta_k \qquad i = 1, 2, \ldots, t$$

By definition of Δ_{k+1}, the theorem is proved. $\qquad\qquad\square$

Theorem 9.13

$$\Delta_k \to \Delta^* \text{ as } k \to \infty.$$

Proof The previous theorem shows that $\{\Delta_k\}$ is a monotonically decreasing sequence, bounded below, and so is convergent to Δ, say. The required result will be proved by assuming that $\Delta > \Delta^*$ and showing that this leads to a contradiction. By this assumption, there exists $\overline{R} = \overline{P}/\overline{Q} \in R_{nm}{}^D$ with $\overline{Q}$ normalized as before, such that

$$\|f - \overline{R}\| = \overline{\Delta} < \Delta.$$

Let δ and M be as in the previous theorem. The string of inequalities leading to (9.15) gives

$$\left| f(x_i) - \frac{P_{k+1}(x_i)}{Q_{k+1}(x_i)} \right| - \Delta_k \leq -\frac{\delta}{M}[\Delta_k - \overline{\Delta}] \frac{Q_k(x_i)}{Q_{k+1}(x_i)} \qquad i = 1, 2, \ldots, t \tag{9.17}$$

and so

$$\Delta_{k+1} - \Delta_k \leq -\frac{\delta}{M}[\Delta - \overline{\Delta}] \frac{Q_k(z_l)}{Q_{k+1}(z_l)} \tag{9.18}$$

where

$$\frac{Q_k(z_l)}{Q_{k+1}(z_l)} \leq \frac{Q_k(x_i)}{Q_{k+1}(x_i)} \qquad i = 1, 2, \ldots, t \tag{9.19}$$

with $z_l = x_j$ for some j, $1 \leqslant j \leqslant t$ and l depending on k. Further, since Δ_k converges to $\Delta > \overline{\Delta}$, (9.18) implies that

$$\lim_{k \to \infty} \frac{Q_k(z_l)}{Q_{k+1}(z_l)} = 0. \tag{9.20}$$

Now from (9.16)

$$Q_{k+1}(x_i) \geqslant Q_k(x_i) \frac{\delta}{M} [\Delta_k - \overline{\Delta}]/\Delta_k \qquad i = 1, 2, \ldots, t$$
$$\geqslant c Q_k(x_i) \qquad\qquad\qquad i = 1, 2, \ldots, t \tag{9.21}$$

provided that

$$c \leqslant \frac{\delta}{M} [\Delta_k - \overline{\Delta}]/\Delta_k. \tag{9.22}$$

Let

$$c = \min \left[\frac{1}{2}, \frac{\delta}{M} [\Delta - \overline{\Delta}]/\Delta \right].$$

Then c is a positive constant which satisfies (9.22) and, using (9.20), there exists an integer k_0 such that

$$\frac{Q_k(z_l)}{Q_{k+1}(z_l)} \leqslant c^t \qquad k \geqslant k_0. \tag{9.23}$$

Using (9.21) and (9.23),

$$\frac{Q_k(z_l)}{Q_{k+1}(z_l)} \prod_{\substack{i=1 \\ i \neq j}}^{t} \frac{Q_k(x_i)}{Q_{k+1}(x_i)} \leqslant c^t \left(\frac{1}{c} \right)^{t-1} = c,$$

and so

$$\prod_{i=1}^{t} \frac{Q_{k+1}(x_i)}{Q_k(x_i)} \geqslant \frac{1}{c} \geqslant 2.$$

It follows that $\prod_{i=1}^{t} Q_k(x_i)$ diverges as $k \to \infty$, which contradicts the assumed normalization and leads to the required result. $\qquad\qquad\square$

Barrodale, Powell, and Roberts (1972) also show that, if the best approximation is normal, then the rate of convergence of the above algorithm is second order. This result is not surprising, in view of the 'Newton's method' nature of the procedure (as indicated in the motivation described earlier). The reader is referred to the paper for details. Each step of the method may be solved as a linear programming problem: the coefficients of $P(x)$ and $Q(x)$ are variables, in addition to a variable w, say, representing the objective function. Thus we require to minimize w subject to the linear constraints

$$[f(x_i) + \Delta_k]Q(x_i) - P(x_i) + Q_k(x_i)w \geqslant 0 \qquad i = 1, 2, \ldots, t$$
$$[-f(x_i) + \Delta_k]Q(x_i) + P(x_i) + Q_k(x_i)w \geqslant 0 \qquad i = 1, 2, \ldots, t$$

and

$$-1 \leqslant q_j \leqslant 1 \qquad j = 0, 1, \ldots, m,$$

where q_j is the coefficient of x^j in $Q(x)$ (verify this). As in the linear problem, it is more efficient to solve the dual problem. Finally, we remark that the algorithm may in theory be applied to the rational problem in $[a, b]$, rather than a discrete subset of it, although the solution of the subproblem at each step is no longer straightforward. Again, second order convergence may be established if the best approximation is normal (Dua and Loeb, 1973).

9.6 The second algorithm of Remes

For approximation on an interval $[a, b]$, the sign alternation property which characterizes the best approximation suggests that a method analogous to the second algorithm of Remes for the linear problem would be appropriate. One obvious difficulty is that the number of points in the alternating set is not known in advance. However, if it is *assumed* that the best approximation is normal, then an algorithm of the required type may be developed. Letting $t = m + n + 2$, at each step of the method points x_i, $i = 1, 2, \ldots, t$, are given in $[a, b]$, and we have to solve the system of nonlinear equations

$$f(x_i) - \frac{P(x_i)}{Q(x_i)} = (-1)^i h \qquad i = 1, 2, \ldots, t$$

for the t unknowns represented by h, and the coefficients of P and Q. The system may be rewritten as

$$(f(x_i) - (-1)^i h)Q(x_i) - P(x_i) = 0 \qquad i = 1, 2, \ldots, t, \qquad (9.24)$$

and the homogeneity of the (unnormalized) form shows that the problem is equivalent to an algebraic eigenvalue problem, where h plays the role of an eigenvalue. Setting

$$P(x) = \sum_{j=0}^{n} p_j x^j$$

$$Q(x) = \sum_{j=0}^{m} q_j x^j$$

it follows that a nontrivial eigenvector of coefficients $p_0, p_1, \ldots, p_n, q_0, q_1, \ldots, q_m$ will exist for each value of h making zero the determinant of the $t \times t$ matrix with ith row

$$[(f(x_i) - (-1)^i h) \ (f(x_i) - (-1)^i h)x_i \ldots (f(x_i) - (-1)^i h)x_i^m \ -1 - x_i \ldots - x_i^n].$$

It is a consequence of this formulation of the problem that there are $(m+1)$ possible values of h, and $(m+1)$ associated rational approximations which satisfy (9.24). In fact Werner (1963) has shown that the eigenvalues are always real. He has also shown that there is at most one value of h such that the corresponding rational function $P(x)/Q(x) \in R_{nm}$; that no such *pole-free* solution (having $Q(x) > 0$ in $[a, b]$) need exist, however, is illustrated by the following example from Maehly (1963).

Example $f(x) = x$ in $[-1, 1]$, $n = 0$, $m = 1$,

$$x_1 = -1, \ x_2 = 0, \ x_3 = 1.$$

The eigenvalues are

$$h_1 = \frac{1}{\sqrt{2}}, \ h_2 = -\frac{1}{\sqrt{2}}$$

with approximations

$$R_1 = \frac{1}{2x - \sqrt{2}} \qquad R_2 = \frac{1}{2x + \sqrt{2}} \qquad \text{(see exercise 21)}.$$

A procedure for the solution of (9.24) as an eigenvalue problem is given by Curtis and Osborne (1966), who use inverse iteration with zero as an initial estimate for the eigenvalue. Even when a pole-free solution exists, however, it does not follow that this is associated with the smallest eigenvalue (exercise 23). The system of nonlinear equations (9.24) may also be solved by more standard techniques (see, for example, Fraser and Hart, 1962; Werner, 1962; Stoer, 1964).

We now state the second algorithm of Remes as it applies to the rational approximation problem. The steps of the method are as follows:

(1) Choose t points in $[a, b]$, say $x_i^{(k)}$, $i = 1, 2, \ldots, t$, with k set to the value 1.

(2) Determine h_k and $R_k = \dfrac{P_k}{Q_k} \in R_{nm}$ to satisfy the system of equations

$$f(x_i^{(k)}) - R_k(x_i^{(k)}) = (-1)^i h_k \qquad i = 1, 2, \ldots, t.$$

(We will assume, without loss of generality, that $h_k \geqslant 0$.)

(3) Determine t local maxima $x_i^{(k+1)}$ of $|f - R_k|$ such that at least one point at which $\|f - R_k\|$ is attained is included, and such that the sign of $f - R_k$ alternates on these points. Unless Theorem 9.9 is satisfied, return to step (2) with $k = k + 1$.

Step (3) can always be executed; see Section 3.5. If step (2) can not, then the method breaks down. Otherwise we can obtain convergence to the best approximation.

Theorem 9.14 Let $f(x) \notin R_{nm}$ and the best approximation to $f(x)$ be normal. Then if there exists $C > 0$ independent of k such that $Q_k(x) \geq C$ for all k, and all $x \in [a, b]$, then

$$h_k \to \min_{R \in R_{nm}} \| f - R \| = h, \text{ as } k \to \infty.$$

Proof Let $\lambda^{(k+1)} \in R^t$ be a nontrivial vector satisfying

$$\sum_{i=1}^{t} \lambda_i^{(k+1)} (x_i^{(k+1)})^j = 0 \qquad j = 0, 1, \ldots, m+n.$$

where, using Lemma 3.1, $\lambda_i^{(k+1)} \neq 0$, $i = 1, 2, \ldots, t$, and the components alternate in sign. Then if

$$\gamma_i^{(k+1)} = \lambda_i^{(k+1)} Q_k(x_i^{(k+1)}) Q_{k+1}(x_i^{(k+1)}) \qquad i = 1, 2, \ldots, t$$

with $\lambda^{(k+1)}$ normalized so that $\sum_{i=1}^{t} |\gamma_i^{(k+1)}| = 1$, it follows that

$$\sum_{i=1}^{t} \gamma_i^{(k+1)} [R_k(x_i^{(k+1)}) - R_{k+1}(x_i^{(k+1)})] = 0. \tag{9.25}$$

Now

$$f(x_i^{(k+1)}) - R_{k+1}(x_i^{(k+1)}) = (-1)^i h_{k+1} \qquad i = 1, 2, \ldots, t$$
$$f(x_i^{(k+1)}) - R_k(x_i^{(k+1)}) = \alpha(-1)^i v_i^{(k)} \qquad i = 1, 2, \ldots, t$$

where $v_i^{(k)} \geq h_k$, $i = 1, 2, \ldots, t$, with strict inequality holding for at least one i, and α is either $+1$ or -1. Thus, using (9.25)

$$h_{k+1} = \sum_{i=1}^{t} |\gamma_i^{(k+1)}| v_i^{(k)},$$

and so

$$h_{k+1} - h_k = \sum_{i=1}^{t} |\gamma_i^{(k+1)}| (v_i^{(k)} - h_k)$$
$$> 0.$$

It follows that $\{h_k\}$ is an increasing sequence, bounded above, and so convergent. We show next that there exists C_1 independent of i, k such that

$$|\gamma_i^{(k+1)}| \geq C_1 > 0 \qquad \text{all } i, k. \tag{9.26}$$

By assumption, this will hold if the components of $\lambda^{(k+1)}$ are similarly bounded away from zero. If they are not, then the points $x_i^{(k+1)}$ cannot remain strictly separated. But these points are, by definition, local maxima of $f - R_k$ with consecutive points having opposite sign. Thus a limiting set must be such that step (2) of the algorithm has a solution with $h_k = 0$, contradicting the fact that successive values of h_k are monotonically increasing. Thus (9.26) holds and so

$$\sum_{i=1}^{t} |\gamma_i^{(k+1)}| (v_i^{(k)} - h_k) \geq C_1 (\| f - R_k \| - h_k).$$

Therefore

$$\| f - R_k \| - h_k \to 0 \text{ as } k \to \infty$$

and since

$$0 \leq h - h_k \leq \| f - R_k \| - h_k$$

the required result follows. ☐

The conditions of this theorem are satisfied provided that the initial set of t points give an approximation $R_1(x)$ which is sufficiently close to the (normal) best approximation (see also Ralston, 1965b; Werner, 1967). The particular choice of initial set is therefore important: Curtis and Osborne (1966) recommend choosing the extrema of the Chebyshev polynomial $T_{t-1}(x)$ (shifted for the interval $[a, b]$); Werner, Stoer, and Bommas (1967) determine $R_1(x)$ as the rational function which interpolates to $f(x)$ at the zeros of this Chebyshev polynomial, and Ralston (1965a) suggests a procedure which may be interpreted as being equivalent to this (Huddleston, 1974). The second algorithm of Remes applied to the rational Chebyshev approximation problem usually works well, and Werner (1962) has shown that second order convergence is possible under reasonable conditions. A result of this kind may be obtained through an analysis similar to that used for the linear case in Theorem 3.12, and this is left as an exercise (see also Curtis and Osborne, 1966).

The method will not work, of course, when the best approximation turns out to be degenerate, and indeed for problems which have solutions which are nearly degenerate, extremely good starting approximations are required. The computation of such approximations should be avoided if possible, particularly as equally good results can be obtained through the use of rational functions having fewer parameters. If a sequence of rational approximations is being computed, having polynomials of increasing degree in both numerator and denominator, then the occurrence of degeneracy can be anticipated (see exercise 28). For a detailed discussion of degeneracy and its consequences, see the papers by Ralston (1965b, 1967, 1973). We remark finally in this context that multiple eigenvalues of the system (9.24) lead to degenerate approximations $P(x)/Q(x)$ (Werner, 1963).

Greater robustness of the second algorithm of Remes may be obtained by combining its merits with those of the differential correction algorithm. Suppose that step (2) of the former method is modified so that we require the solution of a discrete problem on $l \geq m + n + 2$ points in $[a, b]$. If the solution of the continuous problem is not degenerate, then for l sufficiently large, and the points sufficiently well distributed in $[a, b]$, we expect the resulting approximation to be pole-free. Methods based on this idea have been given by Belogus and Liron (1978), and Kaufman, Leeming, and Taylor (1978), in which the differential correction algorithm is used to solve the discrete problems. If the resulting approximation is not pole-free in $[a, b]$, then more points can be added to the discrete set and the process repeated. Numerical results show that such a method can be successful for problems which give difficulties with the traditional Remes algorithm.

Finally, we remark that the method of Remes may be adapted to apply to a problem defined on a discrete set of points. A comparison of eight techniques for such problems is given by Lee and Roberts (1973); a version of the differential correction algorithm comes out best, although the Remes method can be faster when no difficulties arise with degeneracy.

Exercises

23. Show that the pole-free solution to (9.24) is not necessarily associated with the smallest eigenvalue by considering the following example.

$$f(x) = 2 + 2.5x - 0.5x^2 \text{ in } [1, 6], n = 0, m = 1$$
$$x_1 = 1, x_2 = 2, x_3 = 6.$$

24. If $\| Q_k \| = 1$ at each step of the Remes method, prove that if the best approximation R is similarly normalized, under the conditions of Theorem 9.14 we must have $R_k \to R$ as $k \to \infty$.

25. By determining the best approximation from R_{01} on $[-1, \frac{3}{2}]$ to $f(x)$ defined by

$$f(x) = \begin{cases} 2x + 1 & -1 \leqslant x \leqslant \frac{1}{2} \\ 1/x & \frac{1}{2} \leqslant x \leqslant \frac{3}{2} \end{cases}$$

verify that a best approximation on an interval is not necessarily a best approximation on the discrete set defined by the optimal alternating set (Curtis, 1959).

26. Use the second algorithm of Remes to find the best approximation to $\sqrt{x}$ on $[1/16, 1]$ out of R_{01}. Use the initial set $\{1/16, 1/4, 1\}$.

27. Let the best approximation $R(x) \in R_{nm}$ to $f(x) \in C^2[a, b]$ have an alternating set of $t = m + n + 2$ points $a < x_2 < \ldots < x_{t-1} < b$, with no other points in $[a, b]$ at which the maximum error is attained, and assume that

$$Df(a) - DR(a) \neq 0$$
$$Df(b) - DR(b) \neq 0$$
$$D^2 f(x_i) - D^2 R(x_i) \neq 0 \qquad i = 2, 3, \ldots, t-1.$$

By using a method analogous to that used to prove Theorem 3.12, show that the second algorithm of Remes is (to first order terms) equivalent to Newton's method applied to (9.24), and thus can have a second order convergence rate.

28. Let $R_{nm}*(x)$ be the best approximation from R_{nm} to $f(x)$ on $[a, b]$. Then $R_{nm}*(x)$ is degenerate with defect 1 if and only if $R_{n-1, m-1}*(x)$ is nondegenerate and $f(x) - R_{n-1, m-1}*(x)$ has at least $n + m + 1$ alternating extrema (Ralston, 1967).

10

Nonlinear discrete approximation

10.1 Introduction

Most of the problems considered in detail in the previous chapter were very special ones, and use was made of the particular structure of the approximating functions to obtain characterization results and numerical methods for calculating best approximations. Many nonlinear approximation problems, however, lack any directly useful characteristics, and thus there is a need for methods which may be applied in more general situations. Most general methods proceed through the solution of a sequence of linear problems, often of a similar type; a large number of such problems may require to be solved, and thus the ease (or otherwise) with which this can be achieved is an important consideration. We have seen that linear *discrete* problems can often be solved very efficiently, in many important cases in a finite number of steps. Therefore, even when the original nonlinear problem is a continuous one, the treatment of the discrete analogue becomes important. Indeed, in practice there may be no realistic alternative to solving a discretization of the original problem. For this reason, and also because discrete problems arise naturally in many physical situations, for example in the fitting of nonlinear models to discrete data, we devote this final chapter to the treatment of an important general class of nonlinear discrete approximation problems.

Let $\| \, . \, \|$ denote an L_p norm on R^m, where p can take any value between 1 and ∞, *inclusive*, and let $\mathbf{f}(\mathbf{a})$ be a nonlinear mapping of R^n into R^m. Then the problem which we consider is the following:

$$\text{find } \mathbf{a} \in R^n \text{ to minimize } \| \mathbf{f}(\mathbf{a}) \|. \qquad (10.1)$$

Existence of a solution to (10.1) can not, of course, generally be guaranteed. In addition, the characterization of solutions is complicated by the gap between conditions which are necessary and those which are sufficient. However, the necessary conditions which arise as a natural generalization of those occurring in the linear case play a vital role in the analysis of methods for (10.1). To obtain these necessary conditions, it is convenient to introduce the set $V(\mathbf{f})$ defined by

$$V(\mathbf{f}) = \{\mathbf{v} \in R^m, \|\mathbf{f}\| = \mathbf{f}^T \mathbf{v}, \|\mathbf{v}\|^* \leqslant 1\}, \tag{10.2}$$

where $\|.\|^*$ denotes the L_q norm on R^m, with $1/p + 1/q = 1$. The L_p and L_q norms may be readily shown to satisfy the relationship

$$\|\mathbf{u}\| = \max_{\|\mathbf{v}\|^* \leqslant 1} \mathbf{u}^T \mathbf{v} \qquad \mathbf{u}, \mathbf{v} \in R^m \tag{10.3}$$

(see exercise 5 of Chapter 1), where of course the roles of $\|.\|$ and $\|.\|^*$ may be reversed. The set $V(\mathbf{f})$ is therefore just the set of vectors $\mathbf{v}$ at which $\|\mathbf{f}\|$ given by (10.3) is attained.

In what follows, we will use $A = A(\mathbf{a})$ to denote the $m \times n$ matrix with (i, j) element $\partial f_i(\mathbf{a})/\partial a_j$. For convenience, we will permit $\|\mathbf{u}\|$ to be the L_p norm of $\mathbf{u}$, irrespective of the actual dimension of the vector $\mathbf{u}$.

Theorem 10.1   Let $\mathbf{a}$ solve (10.1) and let $\mathbf{f}(\mathbf{a})$ be such that

$$\mathbf{f}(\mathbf{z}) = \mathbf{f}(\mathbf{a}) + A(\mathbf{z} - \mathbf{a}) + O(\|\mathbf{z} - \mathbf{a}\|)$$

for all $\mathbf{z}$ in a neighbourhood of $\mathbf{a}$. Then there exists $\mathbf{v} \in V(\mathbf{f}(\mathbf{a}))$ such that

$$A^T \mathbf{v} = \mathbf{0}.$$

Proof 1   This is an immediate consequence of Theorem 1.6 in the case where $S = S^* = R^m$.

Proof 2   Let $\mathbf{a}$ solve (10.1) and let $\mathbf{f}(\mathbf{a})$ satisfy the condition of the theorem. Define D by

$$D = \{\mathbf{d} \in R^n, \mathbf{d} = A^T \mathbf{v}, \mathbf{v} \in V(\mathbf{f}(\mathbf{a}))\}$$

and assume that $\mathbf{0} \notin D$. Then since D is a closed, bounded convex set in R^n (exercise 1), by Theorem 1.5 there exists $\mathbf{z} \in R^n$ such that for all $\mathbf{d} \in D$,

$$\mathbf{d}^T \mathbf{z} < -\delta \tag{10.4}$$

where δ is a positive number. Now for any $\mathbf{v}(\gamma) \in V(\mathbf{f}(\mathbf{a} + \gamma \mathbf{z}))$, $\gamma > 0$,

$$\begin{aligned}
\|\mathbf{f}(\mathbf{a} + \gamma \mathbf{z})\| &= \mathbf{v}(\gamma)^T \mathbf{f}(\mathbf{a} + \gamma \mathbf{z}) \\
&= \mathbf{v}(\gamma)^T \mathbf{f}(\mathbf{a}) + \gamma \mathbf{v}(\gamma)^T A\mathbf{z} + O(\gamma) \\
&< \|\mathbf{f}(\mathbf{a})\| - \gamma \delta + \gamma (\mathbf{v}(\gamma) - \mathbf{v})^T A\mathbf{z} + O(\gamma)
\end{aligned}$$

for all $\mathbf{v} \in V(\mathbf{f}(\mathbf{a}))$. Let $\mathbf{w}$ be a limit point of $\{\mathbf{v}(\gamma_i)\}$, where $\{\gamma_i\}$ is a sequence of positive values tending to zero. Then we will obtain a contradiction, and therefore the required result, if $\mathbf{w} \in V(\mathbf{f}(\mathbf{a}))$. We have

$$
\begin{aligned}
0 \leqslant \| \mathbf{f}(\mathbf{a}) \| &- \mathbf{v}(\gamma)^T \mathbf{f}(\mathbf{a}) \\
&= \mathbf{v}^T \mathbf{f}(\mathbf{a}) - \mathbf{v}(\gamma)^T \mathbf{f}(\mathbf{a}) \\
&= \mathbf{v}^T \mathbf{f}(\mathbf{a} + \gamma \mathbf{z}) - \mathbf{v}(\gamma)^T \mathbf{f}(\mathbf{a} + \gamma \mathbf{z}) + O(\gamma) \\
&\leqslant O(\gamma)
\end{aligned}
$$

Thus the limit points of $\{\mathbf{v}(\gamma)\}$ lie in $V(\mathbf{f}(\mathbf{a}))$ and the proof is completed. $\qquad \square$

If $\| \mathbf{f}(\mathbf{a}) \|$ is a differentiable function of $\mathbf{a}$ (when $1 < p < \infty$) then the condition of Theorem 10.1 corresponds to a zero derivative (exercise 2). We therefore motivate the following definition.

Definition 10.1 A point $\mathbf{a} \in R^n$ is a *stationary point* of $\| \mathbf{f}(\mathbf{a}) \|$ if there exists $\mathbf{v} \in V(\mathbf{f}(\mathbf{a}))$ such that

$$
A^T \mathbf{v} = \mathbf{0}.
$$

The methods which we consider in this chapter will be shown to be capable of converging to a stationary point of $\| \mathbf{f}(\mathbf{a}) \|$. In general, there is no guarantee that such a point solves (10.1). However, in practice we might expect this to be (at worst) a local minimum, in the sense that $\| \mathbf{f}(\mathbf{z}) \| \geqslant \| \mathbf{f}(\mathbf{a}) \|$ for all points $\mathbf{z}$ in a neighbourhood of $\mathbf{a}$.

Exercises

1. Show that the set D defined above is convex.

2. Obtain the set $V(\mathbf{f})$ defined by (10.2) for all the L_p norms. (Check your results with the sets given in Section 1.5.) For $1 < p < \infty$, show that $V(\mathbf{f})$ is the singleton given by differentiating $\| \mathbf{f} \|$ with respect to the components of $\mathbf{f}$, provided that $\| \mathbf{f} \| \neq 0$.

3. Use the results of the previous exercise to give the necessary conditions appropriate for the L_∞ and L_1 norms in a more convenient. form.

4. Write the nonlinear L_1 and L_∞ problems as mathematical programming problems. Verify that the conditions of Theorem 10.1 are just the Kuhn–Tucker conditions for these problems. (For those who have done a course in optimization.)

5. When $\| \mathbf{f}(\mathbf{a}) \|$ is a convex function of $\mathbf{a}$, prove that the conditions of Theorem 10.1 are also sufficient for $\mathbf{a}$ to solve (10.1).

6. Let $\mathbf{f} \in C^2$ in a neighbourhood of $\mathbf{a} \in R^n$, and let M_k be the $n \times n$ matrix with (i, j) element $\partial^2 f_k(\mathbf{a})/\partial a_i \partial a_j$ and $u_k(\mathbf{d}) = \mathbf{d}^T M_k \mathbf{d}$, $k = 1, 2, \ldots, m$, defined for $\mathbf{d} \in R^n$. If there exists $\mathbf{v} \in V(\mathbf{f}(\mathbf{a}))$ such that $A^T \mathbf{v} = \mathbf{0}$ and $\sum_{k=1}^{m} v_k u_k(\mathbf{d}) > 0$ for all nontrivial $\mathbf{d} \in R^n$, prove that

$$\| \mathbf{f}(\mathbf{a} + \gamma \mathbf{d}) \| > \| \mathbf{f}(\mathbf{a}) \|$$

for all $\mathbf{d} \in R^n$, and all $\gamma > 0$ sufficiently small.

7. If a continuous nonlinear problem is to be solved approximately by discretization, then the technique of computing a sequence of discrete solutions to problems defined on a set of points filling out the interval may not necessarily result in convergence to the best approximation on the interval, even if the latter problem has a unique solution. Illustrate this for the L_∞ norm in $C[0, 1]$ by taking $f_i(\mathbf{a}) = 1/(1 + ax_i)$, $i = 1, 2, \ldots, m$, where $\{x_i\}$ is a subset of $[0, 1]$ not containing 0 (Dunham, 1972). For some general analyses of this problem, see Loeb and Wolfe (1973), Wolfe (1974, 1975, 1977), and Burke (1976).

10.2 Methods of Gauss–Newton type

When the norm is the L_2 norm, a standard approach to (10.1) which has been used for many years is to linearize $\mathbf{f}(\mathbf{a})$ about the current approximation to the solution, and solve a linear L_2 problem. The solution to this problem gives the next approximation and the process is continued. This method most probably dates back to Gauss, and is now known as the Gauss–Newton method. Unfortunately, in this form, it often fails to converge; however, a simple modification can in most cases ensure that convergence to a stationary point is guaranteed. Clearly, the form of the method is such that it may be defined independently of a particular norm, and this fact has been exploited by a number of authors, for example Osborne and Watson (1969) and Cromme (1976) who consider the L_∞ norm, and Osborne and Watson (1971) who treat the L_1 norm. In fact, a convergence analysis may be given for general norms in R^m, although the actual performance of the method depends on the precise nature of the norm (Anderson and Osborne, 1977a; Osborne and Watson, 1978).

We will assume in what follows that there exists a stationary point of $\| \mathbf{f} \|$ in a bounded region B of R^n in which $\mathbf{f}$ is sufficiently smooth that we can write

$$\mathbf{f}(\mathbf{a} + \mathbf{z}) = \mathbf{f}(\mathbf{a}) + A\mathbf{z} + \| \mathbf{z} \|^2 \mathbf{w}(\mathbf{a}, \mathbf{z}) \tag{10.5}$$

where $\mathbf{a} + \gamma \mathbf{z} \in B$, $0 \leqslant \gamma \leqslant 1$, $\| \mathbf{w} \| \leqslant W$ in B, and $A = \nabla \mathbf{f}(\mathbf{a})$. At a point $\mathbf{a} \in R^n$ let

$$\mathbf{r}(\mathbf{h}) = \mathbf{f} + A\mathbf{h}$$

be the form of the linearization of $\mathbf{f}$. Then the problem of finding $\mathbf{h}$ to minimize $\| \mathbf{r} \|$ is frequently referred to as the *linear subproblem*. The modification to the original Gauss–Newton approach is motivated by the following result. We assume that the current approximation is $\mathbf{a}_i$, $\mathbf{f}_i \equiv \mathbf{f}(\mathbf{a}_i)$, $A_i \equiv A(\mathbf{a}_i)$, $\mathbf{h}_i$ solves the linear subproblem, and $\mathbf{r}_i = \mathbf{f}_i + A_i \mathbf{h}_i$.

Theorem 10.2 Let $\mathbf{a}_i \in B$ with $\| \mathbf{r}_i \| < \| \mathbf{f}_i \|$. Then

$$\| \mathbf{f}(\mathbf{a}_i + \gamma \mathbf{h}_i) \| < \| \mathbf{f}_i \|$$

for $\gamma > 0$ sufficiently small.

Proof

$$\begin{aligned}
\mathbf{f}(\mathbf{a}_i + \gamma \mathbf{h}_i) &= \mathbf{f}(\mathbf{a}_i) + \gamma A_i \mathbf{h}_i + \gamma^2 \|\mathbf{h}_i\|^2 \mathbf{w}_i \\
&= \mathbf{f}_i + \gamma(\mathbf{r}_i - \mathbf{f}_i) + \gamma^2 \|\mathbf{h}_i\|^2 \mathbf{w}_i \\
&= \gamma \mathbf{r}_i + (1 - \gamma)\mathbf{f}_i + \gamma^2 \|\mathbf{h}_i\|^2 \mathbf{w}_i.
\end{aligned}$$

Thus if $0 < \gamma \leqslant 1$,

$$\begin{aligned}
\|\mathbf{f}(\mathbf{a}_i + \gamma \mathbf{h}_i)\| &\leqslant \gamma \|\mathbf{r}_i\| + (1 - \gamma)\|\mathbf{f}_i\| + \gamma^2 \|\mathbf{h}_i\|^2 W \\
&= \|\mathbf{f}_i\| + \gamma(\|\mathbf{r}_i\| - \|\mathbf{f}_i\|) + \gamma^2 \|\mathbf{h}_i\|^2 W \qquad (10.6)
\end{aligned}$$

and the result follows. $\qquad\qquad\square$

The fact that $\|\mathbf{r}_i\| < \|\mathbf{f}_i\|$ is required for this result suggests that, since $\|\mathbf{r}_i\| \leqslant \|\mathbf{f}_i\|$ always holds (as can be seen by setting $\mathbf{h}_i = \mathbf{0}$), particular significance is attached to a point $\mathbf{a}_i$ at which $\|\mathbf{r}_i\| = \|\mathbf{f}_i\|$. This is made precise in the following theorem.

Theorem 10.3 $\|\mathbf{r}_i\| = \|\mathbf{f}_i\|$ if and only if $\mathbf{a}_i$ is a stationary point of $\|\mathbf{f}\|$.

Proof Let $\|\mathbf{r}_i\| = \|\mathbf{f}_i\|$. Since $\|\mathbf{r}_i\|$ is a minimum, by Theorem 10.1 there exists $\mathbf{v} \in V(\mathbf{r}_i)$ such that

$$A_i{}^T \mathbf{v} = \mathbf{0}.$$

But $\mathbf{v}^T \mathbf{f}_i = \mathbf{v}^T \mathbf{r}_i = \|\mathbf{r}_i\| = \|\mathbf{f}_i\|$. Thus $\mathbf{v} \in V(\mathbf{f}_i)$ and so $\mathbf{a}_i$ is a stationary point.

Now let $\mathbf{a}_i$ be a stationary point. Then there exists $\mathbf{v} \in V(\mathbf{f}_i)$ such that $A_i{}^T \mathbf{v} = \mathbf{0}$ and so

$$\|\mathbf{f}_i\| = \mathbf{v}^T \mathbf{f}_i = \mathbf{v}^T \mathbf{r}_i \leqslant \|\mathbf{r}_i\|.$$

Since this inequality also goes the other way, the result is proved. $\qquad\square$

Theorem 10.3, and the inequality leading to (10.6), motivates a method for determining an appropriate value of the step length γ, similar to that used in Section 4.3. The inequality to (10.6) gives

$$\frac{\|\mathbf{f}_i\| - \|\mathbf{f}(\mathbf{a}_i + \gamma \mathbf{h}_i)\|}{\gamma(\|\mathbf{f}_i\| - \|\mathbf{r}_i\|)} \geqslant 1 - \frac{\gamma \|\mathbf{h}_i\|^2 W}{\|\mathbf{f}_i\| - \|\mathbf{r}_i\|}. \qquad (10.7)$$

Denoting the left hand side by $\psi(\mathbf{a}_i, \gamma)$, it follows that for fixed i, we have

$$\psi(\mathbf{a}_i, \gamma) \geqslant 1 - \varepsilon(\gamma)$$

where $\varepsilon(\gamma) \to 0$ as $\gamma \to 0$. Thus if γ is chosen to satisfy

$$\psi(\mathbf{a}_i, \gamma) \geqslant \sigma \qquad (10.8)$$

for fixed σ satisfying $0 < \sigma < 1$, this has the effect of ensuring a sufficiently large decrease in the value of $\|\mathbf{f}\|$ and forcing the denominator of $\psi(\mathbf{a}_i, \gamma)$ to zero. We would thus hope to force convergence to a stationary point, and this is confirmed by the following result.

Theorem 10.4 Let $\mathbf{a}_{i+1} = \mathbf{a}_i + \gamma_i \mathbf{h}_i$, where γ_i is chosen as the largest number in the set $\{1, \beta, \beta^2, \ldots\}$, $0 < \beta < 1$, such that (10.8) is satisfied. Then if $\mathbf{a}_0 \in B$, $\|\mathbf{h}_i\| \leqslant H$, all i, $\|\mathbf{f}_i\| - \|\mathbf{r}_i\| \to 0$ as $i \to \infty$.

Proof Assume first that the sequence of values of γ_i is bounded away from zero, by $\hat{\gamma}$ say. Then (10.8) gives

$$\|\mathbf{f}_i\| - \|\mathbf{r}_i\| \leqslant \frac{1}{\hat{\gamma}\sigma} (\|\mathbf{f}_i\| - \|\mathbf{f}_{i+1}\|)$$

$$\to 0 \quad \text{as} \quad i \to \infty$$

since $\{\|\mathbf{f}_i\|\}$ is a decreasing sequence, bounded below.

Now assume that there exists a subsequence of values γ_i (which we do not rename) tending to zero. Then

$$\psi(\mathbf{a}_i, \gamma_i/\beta) < \sigma$$

and so

$$\|\mathbf{f}_i\| - \|\mathbf{r}_i\| < \frac{\gamma_i}{\beta(1-\sigma)} W \|\mathbf{h}_i\|^2.$$

Thus zero is the only limit point of the sequence $\{\|\mathbf{f}_i\| - \|\mathbf{r}_i\|\}$, which must therefore converge to it. $\qquad\Box$

The statement of Theorem 10.4 completes the description of an algorithm for finding a stationary point. It is usual to choose σ as a fairly small positive number, and the value $\sigma = 10^{-4}$ is often suggested. There are various other step length strategies which may be used, but the one given here is convenient from both the theoretical and practical point of view. Notice that the provision of a descent direction does not require that $\|\mathbf{r}\|$ be minimized, but merely that it be reduced to a value less than $\|\mathbf{f}_i\|$. Thus, far from the solution, when the direction $\mathbf{h}_i$ which solves the linear subproblem has no particular merit, some computational savings can be made, particularly if a descent method is being used for the linear subproblem.

For many problems, this modified Gauss–Newton method works well, and is often ultimately convergent with values of γ which can be taken equal to one, when rate of convergence results are possible. In this case we are effectively concerned with the analysis of the method in its original formulation, and it is convenient to consider the linear subproblem in a slightly different form. Following Jittorntrum and Osborne (1979) we define

$$\mathbf{r}(\mathbf{a}, \mathbf{c}) = \mathbf{f}(\mathbf{c}) + A(\mathbf{c})(\mathbf{a} - \mathbf{c}) \tag{10.9}$$

so that the linear subproblem becomes

$$\text{find } \mathbf{a} \in R^n \text{ to minimize } \| \mathbf{r}(\mathbf{a}, \mathbf{c}) \| \qquad (10.10)$$

with $\mathbf{a}_{i+i}$, the $(i + 1)$st approximation, immediately provided as a solution to this problem defined at $\mathbf{c} = \mathbf{a}_i$. Let $\mathbf{a}^*$ be a stationary point of $\| \mathbf{f} \|$, and assume that B is defined by

$$B = \{ \mathbf{a} : \| \mathbf{a} - \mathbf{a}^* \| \leqslant \delta \}$$

in which $\mathbf{f}$ is sufficiently smooth that, in addition to (10.5),

$$\| (A(\mathbf{a}) - A^*)\mathbf{c} \| \leqslant K \| \mathbf{a} - \mathbf{a}^* \| \, \| \mathbf{c} \| \qquad \text{for all } \mathbf{c} \in R^n \qquad (10.11)$$

holds for all $\mathbf{a} \in B$, where $A^* = A(\mathbf{a}^*)$. An important role is played by the concept of strong uniqueness, introduced in Chapter 2. The following result is due essentially to Cromme (1978); the proof is from Jittorntrum and Osborne (1979).

Theorem 10.5 Let the solution to the linear subproblem (10.10) with $\mathbf{c} = \mathbf{a}^*$ be strongly unique. Then if for some j, $\mathbf{a}_j \in B$, with $\| \mathbf{a}_j - \mathbf{a}^* \|$ sufficiently small, the sequence $\{ \mathbf{a}_i \}$ converges to $\mathbf{a}^*$ at a second order rate.

Proof For any $\mathbf{a} \in R^n$, $\mathbf{a}_i \in B$, (10.9) and (10.5) give

$$\mathbf{r}(\mathbf{a}, \mathbf{a}_i) - \mathbf{r}(\mathbf{a}, \mathbf{a}^*) = (A_i - A^*)(\mathbf{a} - \mathbf{a}_i) + \| \mathbf{a}_i - \mathbf{a}^* \|^2 \mathbf{w}(\mathbf{a}, \mathbf{a}_i).$$

Thus

$$\| \mathbf{r}(\mathbf{a}, \mathbf{a}_i) - \mathbf{r}(\mathbf{a}, \mathbf{a}^*) \| \leqslant K \| \mathbf{a}_i - \mathbf{a}^* \| \, \| \mathbf{a} - \mathbf{a}_i \| + W \| \mathbf{a}_i - \mathbf{a}^* \|^2. \qquad (10.12)$$

Let $\mathbf{a} = \mathbf{a}_{i+1}$. Then

$$\| \mathbf{r}(\mathbf{a}_{i+1}, \mathbf{a}_i) - \mathbf{r}(\mathbf{a}_{i+1}, \mathbf{a}^*) \| \leqslant (K + W) \| \mathbf{a}_i - \mathbf{a}^* \|^2 + K \| \mathbf{a}_{i+1} - \mathbf{a}^* \| \, \| \mathbf{a}_i - \mathbf{a}^* \|.$$

By the strong uniqueness assumption, there exists $\gamma > 0$ such that, in particular,

$$\| \mathbf{r}(\mathbf{a}_{i+1}, \mathbf{a}^*) \| \geqslant \| \mathbf{r}(\mathbf{a}^*, \mathbf{a}^*) \| + \gamma \| \mathbf{a}_{i+1} - \mathbf{a}^* \|$$

and so

$$\| \mathbf{r}(\mathbf{a}^*, \mathbf{a}^*) \| + \gamma \| \mathbf{a}_{i+1} - \mathbf{a}^* \| - (K + W) \| \mathbf{a}_i - \mathbf{a}^* \|^2 - K \| \mathbf{a}_{i+1} - \mathbf{a}^* \| \, \| \mathbf{a}_i - \mathbf{a}^* \|$$
$$\leqslant \| \mathbf{r}(\mathbf{a}_{i+1}, \mathbf{a}^*) \| - \| \mathbf{r}(\mathbf{a}_{i+1}, \mathbf{a}_i) - \mathbf{r}(\mathbf{a}_{i+1}, \mathbf{a}^*) \|$$
$$\leqslant \| \mathbf{r}(\mathbf{a}_{i+1}, \mathbf{a}_i) \|$$
$$\leqslant \| \mathbf{r}(\mathbf{a}^*, \mathbf{a}_i) \| \qquad \text{by the definition of } \mathbf{a}_{i+1},$$
$$\leqslant (K + W) \| \mathbf{a}_i - \mathbf{a}^* \|^2 + \| \mathbf{r}(\mathbf{a}^*, \mathbf{a}^*) \|$$

using (10.12) with $\mathbf{a} = \mathbf{a}^*$. Thus

$$\| \mathbf{a}_{i+1} - \mathbf{a}^* \| (\gamma - K \| \mathbf{a}_i - \mathbf{a}^* \|) \leqslant 2(K + W) \| \mathbf{a}_i - \mathbf{a}^* \|^2.$$

Now let $\mathbf{a}_i \in B$ be such that

$$\| \mathbf{a}_i - \mathbf{a}^* \| < \frac{\gamma - \theta}{K}$$

where $\theta > 0$ satisfies $0 < \gamma - \theta < K$. Then

$$\| \mathbf{a}_{i+1} - \mathbf{a}^* \| < \frac{2(K + W)}{\theta} \| \mathbf{a}_i - \mathbf{a}^* \|^2$$

and we have the required result. ☐

Corollary Second order convergence is possible when
  (i) the norm is the L_1 or L_∞ norm and the solution to (10.10) at $\mathbf{a}^*$ is unique, or
  (ii) $\| \mathbf{f}^* \| = 0$ and A^* has rank n.

Proof We have already seen that (i) implies strong uniqueness. For (ii), there exists $\gamma > 0$ such that

$$\| A^*(\mathbf{a} - \mathbf{a}^*) \| \geq \gamma \| \mathbf{a} - \mathbf{a}^* \|$$

for any $\mathbf{a} \in R^n$, so that again the appropriate strong uniqueness result follows. ☐

When the conditions required for this theorem do not hold, the actual rate of convergence may be extremely slow. In particular, if the matrices A_i become rank deficient, $\| \mathbf{h} \|$ may become unbounded, and the method may break down.

Example $m = n = 1, f = 1 + a^2$,

$$r = 1 + a^2 + 2ah$$

so that $|r|$ is minimized when $h = -(1 + a^2)/2a$. Thus $|h| \to \infty$ as $a \to 0$, the solution to the problem.

We turn now to the construction of rather more robust methods, which are free from the kind of restrictions required above, and yet are still capable of possessing satisfactory convergence rates. It is convenient to specialize to particular cases of (10.1), and we consider three important examples.

Exercises

8. Show that a linear discrete L_p approximation problem can be posed as a nonlinear discrete L_2 problem. If $\mathbf{d}$ is the Newton direction at $\mathbf{a}$ for the solution of the normal equations associated with the linear L_p problem, show that the Gauss–Newton direction for the nonlinear L_2 problem is $2(p - 1)\mathbf{d}/p$.

9. Prove Theorem 10.2 by showing that $\mathbf{h}_i$ satisfies (10.4).

10. For the L_p norms, $1 < p < \infty$, how does the role of $\| \mathbf{f}_i \| - \| \mathbf{r}_i \|$ relate to that of the derivative of $\| \mathbf{f}_i \|$ with respect to the components of $\mathbf{a}$ in the suggested step-length strategy?

11. Show that strong uniqueness of the solution to (10.10) with $\mathbf{c} = \mathbf{a}^*$ holds if and only if the point $\mathbf{a}^*$ is a locally strong unique solution to (10.1) in the sense that there exists $\gamma > 0$ such that

$$\| \mathbf{f}(\mathbf{a}) \| \geqslant \| \mathbf{f}(\mathbf{a}^*) \| + \gamma \| \mathbf{a} - \mathbf{a}^* \|$$

for all $\mathbf{a}$ in a neighbourhood of $\mathbf{a}^*$ (Jittorntrum and Osborne, 1979).

12. Second order convergence of the Gauss–Newton method does not require the solutions to the linear subproblems to be unique other than at $\mathbf{c} = \mathbf{a}^*$. Demonstrate this by considering the example in R^3 with the L_∞ norm defined by

$$\mathbf{f} = \begin{bmatrix} \frac{1}{2} + a_1 & + \alpha a_1{}^2 \\ \frac{1}{2} - a_1 & + \beta a_1{}^2 \\ \frac{1}{2} \end{bmatrix} \qquad \text{where } \alpha, \, \beta > 0.$$

(Jittorntrum and Osborne, 1979).

13. The strong uniqueness condition is not *necessary* for second order convergence of the Gauss–Newton method. Illustrate this by considering the example in R^2 with the L_1 norm given by

$$\mathbf{f} = \begin{bmatrix} a_1 + \alpha a_1{}^3 \\ 1 + a_1 - a_1{}^3 \end{bmatrix} \qquad \text{where } \alpha > 0.$$

(Jittorntrum and Osborne, 1979; see also Cromme (1978) for some other aspects of strong uniqueness).

10.3 The L_2 problem

Much attention has been paid to the nonlinear least squares problem, which often arises naturally in discrete data analysis. From Theorem 10.1, and exercise 2, a stationary point of $\| \mathbf{f} \|$ (throughout this section, the norm is the L_2 norm) satisfies

$$A^T \mathbf{f} = \mathbf{0} \tag{10.13}$$

where, as usual in this chapter, $A = \nabla \mathbf{f}$. This is just a system of n nonlinear equations in the n unknown components of $\mathbf{a}$, and so may be solved directly by any method for such problems, for example Newton's method if derivatives of A are available. The Newton direction $\mathbf{d}$ at any point $\mathbf{a}$ is given by

$$\left(A^T A + \sum_{i=1}^{m} f_i G_i \right) \mathbf{d} = - A^T \mathbf{f} \tag{10.14}$$

where G_i is the $n \times n$ Hessian matrix of second partial derivatives of f_i with respect to the components of $\mathbf{a}$. It is clear from this system that the direction at $\mathbf{a}$

generated by the Gauss–Newton method is just that obtained by approximating the second derivatives by zero; alternatively, if the components of **f** become small, then the Newton and Gauss–Newton directions will not differ by much. This suggests that the ultimate convergence of the Gauss–Newton method with γ set to the value 1 will be a consequence of $\| \mathbf{f} \|$ being sufficiently small at the solution (see exercise 14). An analysis of the situation is given by Osborne (1971), who shows also that unless $\| \mathbf{f} \| = 0$ at the solution, the expected rate of convergence is at best first order.

Most methods for the L_2 problem can be interpreted as providing directions of progress through a particular way of approximating second derivatives in the system (10.14). For example, the method of Levenberg (1944) defines a direction **d** as the solution to

$$(A^T A + \lambda^2 I)\mathbf{d} = -A^T \mathbf{f} \tag{10.15}$$

so that **d** may be obtained in a stable manner through the solution of the linear least squares problem:

$$\text{minimize } \| \mathbf{r} \|, \text{ where } \mathbf{r} = \begin{bmatrix} \mathbf{f} \\ \mathbf{0} \end{bmatrix} + \begin{bmatrix} A \\ \lambda I \end{bmatrix} \mathbf{d}.$$

The particular choice $\lambda = 0$ just gives the Gauss–Newton method; for $\lambda > 0$, the matrix $\begin{bmatrix} A \\ \lambda I \end{bmatrix}$ has rank n, and the size of $\| \mathbf{d} \|$ can therefore be controlled. The direction **d** is still a descent direction for $\| \mathbf{f} \|$ (exercise 15), and so the value of λ can in fact be chosen so that a full step in this direction may always be taken, and such that convergence to a stationary point can be guaranteed. An analysis is given by Osborne (1972); algorithms are also given by, for example, More (1978) and Osborne (1976).

The method is now certainly robust, as **d** can always be defined, but convergence may still be extremely slow, particularly if $\| \mathbf{f} \|$ is large at the solution. For such problems, attention has been directed back to closer representations of the system (10.14), so that second derivative information may be incorporated; see, for example, Gill and Murray (1976, 1978). These authors show how approximations **d** to the solution of (10.14) may be obtained in a numerically stable manner, in methods which use either exact second derivatives, or approximations to these using finite differences.

Exercises

14. If $\min_{\| \mathbf{d} \| = 1} \mathbf{d}^T A_i^T A_i \mathbf{d} \geqslant \delta > 0$ for all i, prove that

$$\| \mathbf{h}_i \|^2 \leqslant \frac{1}{\delta}(\| \mathbf{f}_i \|^2 - \| \mathbf{r}_i \|^2).$$

Deduce that the Gauss–Newton method will be convergent with $\gamma = 1$ provided that $\| \mathbf{f}_i \| \leqslant (1 - \sigma)\delta/2W$ for all i.

15. Let $\mathbf{d}$ satisfy

$$(A^TA + C)\mathbf{d} = -A^T\mathbf{f}$$

where C is an $n \times n$ symmetric positive definite matrix. Show that $\mathbf{d}$ may be otained through the solution of a linear least squares problem. Show also that $\mathbf{d}$ is a descent direction for $\|\mathbf{f}\|$ at the current point, if $A^T\mathbf{f} \neq \mathbf{0}$.

16. If $\mathbf{d}$ is defined by (10.15), show that increasing λ^2 effectively biases the Gauss–Newton direction towards the steepest descent direction for $\|\mathbf{f}\|$ at the current point.

10.4 The L_∞ problem

When the norm is the L_∞ norm, we have seen that the Gauss–Newton method is capable of giving second order convergence to a local minimum of $\|\mathbf{f}\|$. If, however, appropriate conditions are not satisfied, and in particular if the iteration is trying to locate a point at which $\mathbf{f}$ attains its norm in fewer than $(n+1)$ points, convergence may be extremely slow, even if the method does not fail (exercise 17). Most of the attempts to improve matters in this case have been concerned in the first instance with modifying the linear subproblem in a manner analogous to that which distinguishes the Levenberg method in the L_2 case, by incorporating a restriction on the size of the solution. Madsen (1975) suggests obtaining the direction $\mathbf{d}$ as the solution to the problem:

$$\text{minimize } \|\mathbf{r}\| \text{ subject to } \|\mathbf{d}\| \leqslant \mu \qquad (10.16)$$

where $\mathbf{r} = \mathbf{f} + A\mathbf{d}$. Values of μ may be chosen so that convergence to a stationary point is obtained (notice that if $\|\mathbf{r}\| < \|\mathbf{f}\|$ then $\mathbf{d}$ is still a descent direction for $\|\mathbf{f}\|$, as Theorem 10.2 may still be applied). An alternative way of achieving a similar effect is given by Anderson and Osborne (1977b), who determine $\mathbf{d}$ by minimizing

$$\max\{\|\mathbf{r}\|, \mu\|\mathbf{d}\|\}$$

again guaranteeing convergence by suitable choice of μ. Both these linear subproblems may be posed and solved as linear programming problems. The merit of this approach again lies in an increase in robustness; however, although also giving some improvement in rate of convergence over the (slowly converging) Gauss–Newton method, the ultimate rate of convergence can still be extremely slow. For such problems, it would appear that, as in the L_2 case, the incorporation of second derivative information is essential for any ultimate recovery of a fast (second order) convergence rate, although the way in which this can be done is not so obvious here. However, the conditions for a stationary point may be written in the form of a system of nonlinear equations to which Newton's method, for example, may be applicable. From exercise 3, these conditions are that there exists

$\lambda \in R^m$ such that

$$A^T \lambda = 0$$
$$\lambda_i \theta_i \geq 0 \qquad i \in \overline{I} \qquad\qquad (10.17)$$
$$\lambda_i = 0 \qquad i \notin \overline{I}$$

where $\overline{I} = \{i : |f_i| = \|\mathbf{f}\|\}$ and $\theta_i = \text{sign}(f_i)$, $i \in \overline{I}$. If the set $\overline{I}$ can be identified, and also the values of θ_i, $i \in \overline{I}$, then the system (10.17), together with a normalization condition on λ and the equations

$$f_i = \theta_i h \qquad i \in \overline{I}$$

is effectively a total of $n + t + 1$ equations (if $\overline{I}$ contains t indices) in the $n + t + 1$ unknowns $\mathbf{a}$, h, and λ_i, $i \in \overline{I}$. If a solution $\mathbf{a}^*$ is such that $h = \|\mathbf{f}^*\|$, and the inequalities $\lambda_i \theta_i \geq 0$, $i \in \overline{I}$, are satisfied, then $\mathbf{a}^*$ is a stationary point. A method based on this approach, which uses Newton's method to solve the nonlinear system, is given by Watson (1979). The method of Anderson and Osborne (1977b) is used to provide the necessary information about $\overline{I}$, together with good initial approximations to the solution. For a similar approach which does not require explicit second derivatives, see Hald and Madsen (1978).

Exercises

17. If the Gauss–Newton method applied to (10.1) with the L_∞ norm is trying to locate a point $\mathbf{a}^*$ for which $\mathbf{f}^*$ has fewer than $(n + 1)$ components which (in modulus) attain $\|\mathbf{f}^*\|$, show that the rate of convergence cannot be second order.

18. By writing the linear subproblem (10.16) as a linear programming problem, and formulating the dual problem, prove that necessary and sufficient conditions that $\mathbf{d} \in R^n$ solves (10.16) are that there exist vectors $\mathbf{v} \in V(\mathbf{r}(\mathbf{d}))$, $\mathbf{w} \in V(\mathbf{d})$ and a scalar $\alpha \geq 0$ such that

$$\alpha(\mu - \|\mathbf{d}\|) = 0$$
$$A^T \mathbf{v} + \alpha \mathbf{w} = 0$$

where V is given by (10.2) for the L_∞ norm.

10.5 The L_1 problem

When the norm in (10.1) is the L_1 norm, bounded step methods are again appropriate to give some improvement over the basic Gauss–Newton method when that method is slowly convergent. Anderson and Osborne (1977b) minimize

$$\max\{\|\mathbf{r}\|_1, \mu\|\mathbf{d}\|_\infty\}$$

by a linear programming method. An alternative approach is to solve the subproblem:

$$\text{minimize } \|\mathbf{r}\|_1 \text{ subject to } \|\mathbf{d}\|_\infty \leq \mu \qquad (10.18)$$

which can be more efficiently solved, for example by the method of Barrodale and Roberts (1978). Algorithms based on this linear subproblem are given by McLean and Watson (1980). As in the L_∞ case, it is possible to incorporate second derivative information by direct solution of the necessary conditions. From exercise 3, these correspond to the existence of numbers v_i such that

$$\sum_{i \in Z} v_i \alpha_i = - \sum_{i \notin Z} \theta_i \alpha_i \qquad (10.19)$$

where $Z = \{i : f_i = 0\}$, $\theta_i = \text{sign}(f_i)$, $i \notin Z$, and α_i^T denotes the ith row of A. If the set Z, and signs θ_i, $i \notin Z$, can be identified, then (10.19), together with

$$f_i(\mathbf{a}) = 0 \qquad i \in Z,$$

give a total of $n + t$ equations (where Z contains t indices) in the $n + t$ unknows $\mathbf{a}$, $v_i, i \in Z$. A method based on this approach, which uses the above bounded variable linear subproblem method to obtain information about Z and θ_i, $i \notin Z$, and Newton's method to solve the nonlinear system, is described by McLean and Watson (1980).

A completely different approach to the nonlinear L_1 problem is taken by El-Attar, Vidyasagar, and Datta (1979). They proceed by minimizing the differentiable function

$$P(\mathbf{a}, \varepsilon) = \sum_{i=1}^{m} (f_i^2(\mathbf{a}) + \varepsilon)^{\frac{1}{2}},$$

for a sequence of positive values of ε tending to zero. Finally, a descent method analogous to that of Section 6.4 has been developed by Bartels and Conn (to appear).

Exercises

19. If the Gauss–Newton method applied to (10.1) with the L_1 norm is trying to locate a point $\mathbf{a}^*$ with $\mathbf{f}^*$ having fewer than n zero components, show that the rate of convergence cannot be second order.

20. Let $V^{(1)}(\mathbf{f})$ and $V^{(\infty)}(\mathbf{f})$ denote the sets (10.2) for the L_1 and L_∞ norms respectively. By writing the linear subproblem (10.18) as a linear programming problem, and formulating the dual problem, prove that necessary and sufficient conditions for $\mathbf{d} \in R^n$ to be a solution to (10.18) are that there exist vectors $\mathbf{v} \in V^{(1)}(\mathbf{r}(\mathbf{d}))$, $\mathbf{w} \in V^{(\infty)}(\mathbf{d})$, and a scalar $\lambda \geqslant 0$ such that

$$\lambda(\mu - \|\mathbf{d}\|_\infty) = 0$$
$$A^T \mathbf{v} + \lambda \mathbf{w} = \mathbf{0}.$$

References

Anderson, D. H., and Osborne, M. R. (1976). Discrete, linear approximation problems in polyhedral norms, *Num. Math.*, **26**, 179–189.

Anderson, D. H., and Osborne, M. R. (1977a). Discrete, nonlinear approximation problems in polyhedral norms, *Num. Math.*, **28**, 143–156.

Anderson, D. H., and Osborne, M. R. (1977b). Discrete, nonlinear approximation problems in polyhedral norms: a Levenberg-like algorithm, *Num. Math.*, **28**, 157–170.

Andreassen, D. O., and Watson, G. A. (1976). Linear Chebyshev approximation without Chebyshev sets, *B.I.T.*, **16**, 349–362.

Ascher, U. (1976). *Linear programming algorithms for the Chebyshev solution to a system of consistent linear equations*, University of Wisconsin MRC Report 1617.

Ascher, U. (1978). On the invariance of the interpolation points of the discrete l_1 approximation, *J. Approx. Th.*, **24**, 83–91.

Bak, J. (1977). On the efficiency of general rational approximation, *J. Approx. Th.*, **20**, 46–50.

Bak, J., and Newman, D. J. (1972). Muntz–Jackson theorems in $L^p[0, 1]$ and $C[0, 1]$, *Amer. J. Math.*, **94**, 437–457.

Bak, J., and Newman, D. J. (1978). Rational combinations of x^{λ_k}, $\lambda_k \geq 0$ are always dense in $C[0, 1]$, *J. Approx. Th.*, **23**, 155–157.

Barrodale, I. (1968). L_1 approximation and the analysis of data, *Appl. Stat.*, **17**, 51–57.

Barrodale, I. (1970). On computing best L_1 approximations. In A. Talbot (Ed.), *Approximation Theory*, Academic Press, London.

Barrodale, I., and Phillips, C. (1975a). An improved algorithm for discrete Chebyshev linear approximation. In B. L. Hartnell and H. C. Williams (Eds), *Proc. Fourth Manitoba Conf. on Numerical Mathematics*, Utilitas Math. Pub. Co.

Barrodale, I., and Phillips, C. (1975b). Algorithm 495: Solution of an overdetermined system of linear equations in the Chebyshev norm, *A.C.M. Trans. Math. Software*, **1**, 264–270.

Barrodale, I., Powell, M. J. D., and Roberts, F. D. K. (1972). The differential correction algorithm for rational L_∞ approximation, *SIAM J. Num. Anal.*, **9**, 493–504.

Barrodale, I., and Roberts, F. D. K. (1972). *An improved algorithm for discrete l_1 linear approximation*, University of Wisconsin MRC Report 1172.

Barrodale, I., and Roberts, F. D. K. (1973). An improved algorithm for discrete l_1 linear approximation, *SIAM J. Num. Anal.*, **10**, 839–848.

Barrodale, I., and Roberts, F. D. K. (1978). An efficient algorithm for discrete L_1 linear approximation with linear constraints, *SIAM J. Num. Anal.*, **15**, 603–611.

Barrodale, I., and Young, A. (1966). Algorithms for best L_1 and L_∞ linear approximation on a discrete set, *Num. Math.*, **8**, 295–306.

Bartels, R. H. and Conn, A. R. (to appear). An exact penalty algorithm for solving the nonlinear l_1 problem.

Bartels, R. H., Conn, A. R., and Charalambous, C. (1978). On Cline's direct method for solving overdetermined linear systems in the l_∞ sense, *SIAM J. Num. Anal.*, **15**, 255–270.

Bartels, R. H., Conn, A. R., and Sinclair, J. W. (1978). Minimisation techniques for piecewise differentiable functions: the l_1 solution to an overdetermined linear system, *SIAM J. Num. Anal.*, **15**, 224–241.

Bartels, R. H., and Golub, G. H. (1968). Stable numerical methods for obtaining the Chebyshev solution to an overdetermined system of equations, *Comm. A.C.M.*, **11**, 401–406.

Bartelt, M., and McLaughlin, H. W. (1973). Characterisation of strong unicity in approximation theory, *J. Approx. Th.*, **9**, 255–266.

Belogus, D., and Liron, N. (1978). DCR 2: An improved algorithm for l_∞ rational approximation on intervals, *Num. Math.*, **31**, 17–29.

Bernstein, S. N. (1912). Sur l'ordre de la meilleure approximation des fonctions continues par les polynômes de degré donné, *Mem. Acad. Roy. Belg.*, **4**, 1–104.

Bernstein, S. N. (1938). Sur le problème inverse de la théorie de la meilleure approximation des fonctions continues, *Comptes Rendues*, **206**, 1520–1523.

Birkhoff, G., and Priver, A. (1967). Hermite interpolation errors for derivatives, *J. Math. and Physics*, **46**, 440–447.

Boehm, B. W. (1965). Existence of best rational Tchebycheff approximations, *Pacific J. of Math.*, **15**, 19–28.

Boggs, P. T. (1974). A new algorithm for the Chebyshev solution of overdetermined linear systems, *Maths. of Comp.*, **28**, 203–217.

de Boor, C. (1963). Best approximation properties of spline functions of odd degree, *J. Math. Mech.*, **12**, 747–749.

de Boor, C. (1972). On calculating with B-splines, *J. Approx. Th.*, **6**, 50–62.

de Boor, C. (1973). Good approximation by splines with variable knots. In A. Meir and A. Sharma (Eds), *Spline Functions and Approximation Theory*, Birkhauser-Verlag, Basel.

de Boor, C. (1977). Package for calculating with B-splines, *SIAM J. Num. Anal.*, **14**, 441–472.

de Boor, C. (1978). *A Practical Guide to Splines*, Springer-Verlag, New York.

de Boor, C., and Fix, G. J. (1973). Spline approximation by quasi-interpolants, *J. Approx. Th.*, **8**, 19–45.

de Boor, C., Lyche, T., and Schumaker, L. (1976). On calculating with B-splines II: Integration. In *Proc. Conf. Oberwolfach*, ISNM 30, Birkhauser-Verlag, Basel.

de Boor, C., and Rice, J. R. (1968). *Least squares cubic spline approximation II—variable knots*, Purdue University Report CSDTR21.

Braess, D. (1967). Approximation mit Exponentialsummen, *Computing*, **2**, 309–321.

Braess, D. (1970). Die Konstruktion der Tschebyscheff-Approximierenden bei der Anpassung mit Exponentialsummen, *J. Approx. Th.*, **3**, 261–273.

Braess, D. (1971). Chebyshev approximation by spline functions with free knots, *Num. Math.*, **17**, 357–366.

Braess, D. (1973). Chebyshev approximation by γ-polynomials, *J. Approx. Th.*, **9**, 20–43.

Braess, D. (1974a). Chebyshev approximation by γ-polynomials II, *J. Approx. Th.*, **11**, 16–37.

Braess, D. (1974b). Geometrical characterisations for nonlinear uniform approximation, *J. Approx. Th.*, **11**, 260–274.

Buck, R. C. (1968). Alternation theorems for functions of several variables, *J. Approx. Th.*, **1**, 325–334.

Burchard, H. G. (1974). Splines (with optimal knots) are better, *Applic. Anal.*, **3**, 309.

Burgoyne, F. D. (1967). Practical L_p polynomial approximation, *Maths. of Comp.*, **21**, 113–115.

Burke, M. E. (1976). Nonlinear best approximation on discrete sets, *J. Approx. Th.*, **16**, 133–141.

Butterfield, K. R. (1976). The computation of all the derivatives of a B-spline basis, *JIMA*, **17**, 15–25.

Cadzow, J. A. (1973). A finite algorithm for the minimum l_∞ solution to a system of consistent linear equations, *SIAM J. Num. Anal.*, **10**, 607–617.

Cadzow, J. A. (1974). An efficient algorithmic procedure for obtaining a minimum l_∞ norm solution to a system of consistent linear equations, *SIAM J. Num. Anal.*, **11**, 1151–1165.

Carlson, R. E., and Hall, C. A. (1973). Error bounds for bicubic spline interpolation, *J. Approx. Th.*, **7**, 41–47.

Carroll, M. P., and McLaughlin H. W. (1973). On L^1 approximation of discontinuous functions, *J. Approx. Th.*, **8**, 129–132.

Chalmers, B. L., and Taylor, G. D. (1978). *Uniform approximation with constraints*, preprint.

Cheney, E. W. (1966). *Introduction to Approximation Theory*, McGraw Hill, New York.

Cheney, E. W., and Loeb, H. L. (1961). Two new algorithms for rational approximation, *Num. Math.*, **3**, 72–75.

Cheney, E. W., and Loeb, H. L. (1962). On rational Chebyshev approximation, *Num. Math.*, **4**, 124–127.

Cheney, E. W., and Loeb, H. L. (1964). Generalised rational approximation, *SIAM J. Num. Anal.*, **1**, 11–25.

Cheney, E. W., and Southard, T. H. (1963). A survey of methods for rational approximation, with particular reference to a new method based on a formula of Darboux, *SIAM Rev.*, **5**, 219–231.

Claerbout, J. F., and Muir, F. (1973). Robust modelling with erratic data, Geophysics, **38**, 826–844.

Cline, A. K. (1972). Rate of convergence of Lawson's algorithm, *Maths. of Comp.*, **26**, 167–176.

Cline, A. K. (1976). A descent method for the uniform solution to overdetermined systems of linear equations, *SIAM J. Num. Anal.*, **13**, 293–309.

Collatz, L. (1956). Approximationen von Funktionen bei einer und bei mehreren unabhangigen Veränderlichen, *Z. ang. Math. u. Mech.*, **36**, 198–211.

Cox, M. G. (1972). The numerical evaluation of B-splines, *JIMA*, **10**, 134–149.

Cox, M. G. (1975). An algorithm for spline interpolation, *JIMA*, **15**, 95–108.

Cox, M. G. (1977). A survey of numerical methods for data and function approximation. In D. Jacobs (Ed.), *The State of the Art in Numerical Analysis*, Academic Press, London.

Cox, M. G. (1978a). The incorporation of boundary conditions in spline approximation problems. In G. A. Watson (Ed.), *Numerical Analysis*, Dundee 1977, Springer-Verlag, Berlin.

Cox, M. G. (1978b). The numerical evaluation of a spline from its B-spline representation, *JIMA*, **21**, 135–143.

Cox, M. G., and Hayes, J. G. (1973). *Curve fitting: a guide and suite of algorithms for the non-specialist user*, Report NAC 26, N.P.L.

Cromme, L. (1976). Eine Klasse von Verfahren zur Ermittlung bester nicht-linearer Tschebyscheff-Approximationen, *Num. Math.*, **25**, 447–459.

Cromme, L. (1978). Strong uniqueness. A far-reaching criterion for the convergence analysis of iterative processes, *Num. Math.*, **29**, 179–193.

Curry, H. B., and Schoenberg, I. J. (1947). On spline distributions and their limits: the Polya distributions, Abstract, *Bull. A.M.S.*, **53**, 114.

Curry, H. B., and Schoenberg, I. J. (1966). On Polya frequency functions IV. The fundamental spline functions and their limits, *J. Analyse Math.*, **17**, 71–107.

Curtis, P. C. (1959). *N*-parameter families and best approximation, *Pac. J. of Math.*, **9**, 1013–1027.

Curtis, A. R. (1970). The approximation of a function of one variable by cubic splines. In

220

J. G. Hayes (Ed.), *Numerical Approximation to Functions and Data*, Athlone Press, London.

Curtis, A. R., and Osborne, M. R. (1966). The construction of minimax rational approximations to functions, *Comp. J.*, **9**, 286–293.

Descloux; J. (1963). Approximations in L^p and Chebyshev approximations, *SIAM J.*, **11**, 1017–1026.

Devore, R. A. (1974). A property of Chebyshev polynomials, *J. Approx. Th.*, **12**, 418–419.

Dua, S. N., and Loeb, H. L. (1973). Further remarks on the differential correction algorithm, *SIAM J. Num. Anal.*, **10**, 123–126.

Dunham, C. B. (1972). Minimax nonlinear approximation by approximation on subsets, *Comm. A.C.M.*, **15**, 351.

Dunham, C. B. (1975). Uniqueness of best Chebyshev approximation on subsets, *J. Approx. Th.*, **14**, 148–151.

Duris, C. S., and Sreedharan, V. P. (1968). Chebyshev and l_1 solutions of linear equations using least squares solutions, *SIAM J. Num. Anal.*, **5**, 491–505.

Duris, C. S., and Temple, M. G. (1973). A finite step algorithm for determining the 'strict' Chebyshev solution to $Ax = b$, *SIAM J. Num. Anal.*, **10**, 690–699.

Ekblom, H. (1973). Calculation of linear best L_p approximations, *BIT*, **13**, 292–300.

El-Attar, R. A., Vidyasagar, M., and Datta, S. R. K. (1979). An algorithm for l_1-norm minimisation with applications to nonlinear l_1-approximation, *SIAM J. Num. Anal.*, **16**, 70–86.

Elosser, P. D. (1978). Approximation of powers of x by polynomials, *J. Approx. Th.*, **23**, 163–174.

Faber, G. (1914). Über die interpolatorische Darstellung stetiger Funktionen, *Deutsche Math. Jahr.*, **23**, 192–210.

Fletcher, R., Grant, J. A., and Hebden, M. D. (1971). The calculation of linear best L_p approximations, *Comp. J.*, **14**, 276–279.

Fletcher, R., Grant, J. A., and Hebden, M. D. (1974a). Minimax approximation as the limit of best L_p approximation, *SIAM J. Num. Anal.*, **11**, 123–136.

Fletcher, R., Grant, J. A., and Hebden, M. D. (1974b). The continuity and differentiability of the parameters of best linear L_p approximations, *J. Approx. Th.*, **10**, 69–73.

Fox, L., and Parker, I. B. (1968). *Chebyshev Polynomials in Numerical Analysis*, Oxford University Press, London.

Fraser, W., and Hart, J. F. (1962). On the computation of rational approximations to continuous functions, *Comm. A.C.M.*, **5**, 401–403.

Freud, G. (1958). Eine Ungleichung für Tschebyscheffsche Approximations-polynome, *Acta. Scient. Math.*, **19**, 162–164.

Gaffney, P. W. (1976). The calculation of indefinite integrals of B-splines, *JIMA*, **17**, 37–42.

Gearhart, W. B. (1973). Some Chebyshev approximations by polynomials in two variables, *J. Approx. Th.*, **8**, 195–209.

Gill, P. E., and Murray, W. (1976). Nonlinear least squares and nonlinearly constrained optimization. In G. A. Watson (Ed.), *Numerical Analysis, Dundee 1975*, Springer-Verlag, Berlin.

Gill, P. E., and Murray, W. (1978). Algorithms for the solution of the nonlinear least-squares problem, *SIAM J. Num. Anal.*, **15**, 977–992.

Glashoff, K., and Schultz, R. (1979). Über die genaue Berechnung von besten L^1-Approximierenden, *J. Approx. Th.*, **25**, 280–293.

Goldstein, A. A., and Price, J. F. (1967). An effective algorithm for minimization, *Num. Math.*, **10**, 184–189.

Golitschek, M. V. (1976). Jackson–Muntz–Szasz Theorems in $L^p[0, 1]$ and $C[0, 1]$ for complex exponents, *J. Approx. Th.*, **18**, 13–29.

Golub, G. H. (1965). Numerical methods for solving linear least squares problems, *Num. Math.*, **7**, 206–216.

Golub, G. H., and Smith, L. B. (1971). Algorithm 414: Chebyshev approximation of continuous functions by a Chebyshev system, *Comm. A.C.M.*, **14**, 737–746.

Gourlay, A. R., and Watson, G. A. (1973). *Computational Methods for Matrix Eigenproblems*, Wiley, Chichester.

Haar, A. (1918). Die Minkowskische Geometrie und die Annäherung an stetige Funktionen, *Math. Ann.*, **78**, 294–311.

Hadley, G. (1962). *Linear Programming*, Addison-Wesley, Reading, Mass.

Hald, J., and Madsen, K. (1978). *A 2-stage algorithm for minimax optimization*, Report NI–78–11, Technical University of Denmark.

Hall, C. A. (1968). On error bounds for spline interpolation, *J. Approx. Th.*, **1**, 209–218.

Hall, C. A. (1973). Natural cubic and bicubic spline interpolation, *SIAM J. Num. Anal.*, **10**, 1055–1060.

Hardy, G. H., Littlewood, J. E., and Polya, G. (1934). *Inequalities*, Cambridge University Press, London.

Hayes, J. G. (1970). Curve fitting by polynomials in one variable. In J. G. Hayes (Ed.), *Numerical Approximation to Functions and Data*, Athlone Press, London.

Hayes, J. G., and Halliday, J. (1974). The least squares fitting of cubic spline surfaces to general data sets, *JIMA*, **14**, 89–103.

Hebden, M. D. (1971). A bound on the difference between the Chebyshev norm and the Holder norms of a function, *SIAM J. Num. Anal.*, **8**, 270–277.

Henry, M. S., and Weinstein, S. E. (1974). Best rational product approximation of functions II, *J. Approx. Th.*, **12**, 6–22.

Hobby, C. R., and Rice, J. R. (1965). A moment problem in L_1 approximation, *Proc. A.M.S.*, **16**, 665–670.

Hobby, C. R., and Rice, J. R. (1967). Approximation from a curve of functions, *Arch. Rat. Mech. Anal.*, **27**, 91–106.

Holmes, R. B. (1975). *Geometric Functional Analysis and its Applications*, Springer-Verlag, New York.

Hopper, M. J., and Powell, M. J. D. (1977). A technique that gains speed and accuracy in the minimax solution of overdetermined linear equations. In J. R. Rice (Ed.), *Mathematical Software III*, Academic Press, New York.

Huddleston, R. E. (1974). On the conditional equivalence of two starting methods for the second algorithm of Remes, *Maths. of Comp.*, **28**, 569–572.

Jackson, D. (1921). Note on a class of polynomials of approximation, *Trans. A.M.S.*, **22**, 320–326.

Jittorntrum, K., and Osborne, M. R. (1979). *Strong uniqueness and second order convergence in nonlinear discrete approximation*, preprint.

Jones, R. C., and Karlowitz, L. A. (1970). Equioscillation and nonuniqueness in the approximation of continuous functions, *J. Approx. Th.*, **3**, 138–145.

Jupp, D. L. B. (1978). Approximation to data by splines with free knots, *SIAM J. Num. Anal.*, **15**, 328–343.

Kahng, S. W. (1972). Best L_p approximation, *Maths. of Comp.*, **26**, 505–508.

Kammler, D. W. (1976). An alternation characterization of best uniform approximations on noncompact intervals, *J. Approx. Th.*, **16**, 97–104.

Karlin, S. (1971). Best quadrature formulas and splines, *J. Approx. Th.*, **4**, 59–90.

Karlin, S., and Karon, J. M. (1972). On Hermite–Birkhoff interpolation, *J. Approx. Th.*, **6**, 90–114.

Karlin, S., and Studden, W. J. (1966). *Tchebycheff Systems: With Applications in Analysis and Statistics*, Interscience, New York.

Karlovitz, L. (1978). Remarks on the construction of Schauder bases and on the Polya algorithm. In D. C. Handscomb (Ed.), *Multivariate Approximation*, Academic Press, London.

Karon, J. M. (1978). Computing improved Chebyshev approximations by the continuation method. I: Description of an algorithm. *SIAM J. Num. Anal.*, **15**, 1269–1288.

Karon, J. M., and Starner, J. W. (to appear). Computing improved Chebyshev approximations by the continuation method. II: Numerical results.

Kaufman, E. H., Leeming, D. J., and Taylor, G. D. (1978). A combined Remes–Differential Correction algorithm for rational approximation, *Maths. of Comp.*, **32**, 233–242.

Kelley, J. E. (Jr.) (1958). An application of linear programming to curve fitting, *SIAM J.*, **6**, 15–22.

Kershaw, D. (1971). A note on the convergence of interpolatory cubic splines, *SIAM J. Num. Anal.*, **8**, 67–74.

Kirchberger, P. (1903). Uber Tchebychefsche Annaherungsmethoden, *Math. Ann.*, **57**, 509–540.

Krabs, W. (1967). Uber differenzierbare asymptotische konvexe Funktionenfamilien bei der nicht–linearen gleichmaßigen Approximation, *Arch. Rat. Mech. Anal.*, **27**, 275–288.

Kripke, B., and Rivlin, T. J. (1965). Approximation in the metric of $L^1(x, \mu)$, *Trans. A.M.S.*, **119**, 101–122.

Laplace, P. S. (1799). *Mécanique Céleste*, Tome 111, No. 39.

Lawson, C. L., and Hanson, R. J. (1974). *Solving Least Squares Problems*, Prentice Hall, Englewood Cliffs, New Jersey.

Lee, C. M., and Roberts, F. D. K. (1973). A comparison of algorithms for rational approximation, *Maths. of Comp.*, **27**, 111–121.

Levenberg, K. (1944). A method for the solution of certain non-linear problems in least squares, *Quart. J. Appl. Math.*, **2**, 164–168.

Leviatan, D. (1974). On the Jackson–Muntz theorem, *J. Approx. Th.*, **10**, 1–5.

Levitan, M. L., and Lynn, R. Y. S. (1976). An overdetermined linear system, *J. Approx. Th.*, **18**, 264–277.

Loeb, H. L., and Wolfe, J. M. (1973). Discrete nonlinear approximation, *J. Approx. Th.*, **7**, 365–385.

Lorentz, G. G. (1974). The Birkhoff interpolation problem: new methods and results. In *Proc. Conf. Oberwolfach*, ISNM **25**, Birkhauser–Verlag.

Lyche, T. (1978). A note on the condition numbers of the B-spline basis, *J. Approx. Th.*, **22**, 202–205.

Lyche, T., and Schumaker, L. (1973). Computation of smoothing and interpolating splines via local bases, *SIAM J. Num. Anal.*, **10**, 1027–1038.

Madsen, K. (1975). An algorithm for minimax solution of over-determined systems of non-linear equations, *JIMA*, **16**, 321–328.

Maehly, H. J. (1963). Methods of fitting rational approximations, Part II, *JACM*, **10**, 257–266.

Mairhuber, J. C. (1956). On Haar's theorem concerning Chebyshev approximation problems having unique solutions, *Proc. A.M.S.*, **7**, 609–615.

Marti, J. T. (1975). A method for the numerical computation of best L_1 approximations of continuous functions. In *Proc. Conf. Oberwolfach*, ISNM **26**, Birkhauser–Verlag.

McLaughlin, H. W., and Somers, K. B. (1975). Another characterization of Haar subspaces, *J. Approx. Th.*, **14**, 93–102.

McLean, R. A., and Watson, G. A. (1980). Numerical methods for nonlinear discrete L_1 approximation problems. In *Proc. Conf. Oberwolfach*, ISNM, Birkhauser–Verlag.

Meinardus, G. (1967). *Approximation of Functions: Theory and Numerical Methods*, Springer–Verlag, Berlin.

Meinardus, G., and Schwedt, D. (1964). Nicht-lineare Approximationen, *Arch. Rat. Mech. Anal.*, **17**, 297–326.

Meinardus, G., and Taylor, G. D. (1976). Lower estimates for the error of best uniform approximation, *J. Approx. Th.*, **16**, 150–161.

Merle, G., and Spath, H. (1974). Computational experience with discrete L_p approximation, *Computing*, **12**, 315–321.

Micchelli, C. A., and Rivlin, T. J. (1974). Some new characterizations of the Chebyshev polynomials, *J. Approx. Th.*, **12**, 420–424.

Mitchell, A. R., and Wait, R. (1977). *The Finite Element Method in Partial Differential Equations*, Wiley, London.

Moré, J. J. (1978). The Levenberg–Marquardt algorithm: implementation and theory. In G. A. Watson (Ed.), *Numerical Analysis*, Dundee 1977, Springer–Verlag, Berlin.

Motzkin, T. S. (1949). Approximation by curves of a unisolvent family, *Bull. A. M. S.*, **55**, 789–793.

Motzkin, T. S., and Walsh, J. L. (1956). Least pth power polynomials on a finite point set, *Trans. A.M.S.*, **83**, 371–396.

Newman, D. J. (1964). Rational approximation to $|x|$, *Michigan Math. J.*, **11**, 11–14.

Newman, D. J., and Shapiro, H. S. (1962). Some theorems on Čebyšev approximation, *Duke Math. J.*, **30**, 673–682.

Osborne, M. R. (1971). An algorithm for discrete, nonlinear best approximation problems. In *Proc. Conf. Oberwolfach*, ISNM **16**, Birkhauser-Verlag.

Osborne, M. R. (1972). Some aspects of nonlinear least squares calculations. In F. A. Lootsma (Ed.), *Numerical Methods for Non-linear Optimization*, Academic Press, London.

Osborne, M. R. (1976). Nonlinear least squares: the Levenberg algorithm revisited, *J. Aust. Math. Soc., Ser. B*, **19**, 343–357.

Osborne, M. R., and Watson, G. A. (1967). On the best linear Chebyshev approximation, *Comp. J.*, **10**, 172–177.

Osborne, M. R., and Watson, G. A. (1969). An algorithm for minimax approximation in the non-linear case, *Comp. J.*, **12**, 64–69.

Osborne, M. R., and Watson, G. A. (1971). On an algorithm for non-linear L_1 approximation, *Comp. J.*, **14**, 184–188.

Osborne, M. R., and Watson, G. A. (1978). Nonlinear approximation problems in vector norms. In G. A. Watson (Ed.), *Numerical Analysis*, Dundee 1977, Springer–Verlag, Berlin.

Phillips, G. M. (1968). Estimate of the maximum error in best polynomial approximations, *Comp. J.*, **11**, 110–111.

Phillips, G. M. (1970). Error estimates for best polynomial approximation. In A. Talbot (Ed.), *Approximation Theory*, Academic Press, London.

Polya, G. (1913). Sur une algorithme toujours convergent pour obtenir les polynomes de meilleure approximation de Tchebysheff pour une fonction continue quelconque, *Comptes Rendues*, **157**, 840–843.

Powell, M. J. D. (1967). On the maximum errors of polynomial approximations defined by interpolation and by least squares criteria, *Comp. J.*, **9**, 404–407.

Powell, M. J. D. (1970). Curve fitting by splines in one variable. In J. G. Hayes (Ed.), *Numerical Approximation to Functions and Data*, Athlone Press, London.

Powell, M. J. D. (1972). *A Fortran subroutine for calculating a cubic spline approximation to a given function*, Report No. R 7308, AERE Harwell.

Rademacher, H., and Schoenberg, I. J. (1950). Helly's theorem on convex domains and Tchebycheff's approximation problem, *Can J. of Math.*, **2**, 245–256.

Ralston, A. (1965a). *A First Course in Numerical Analysis*, McGraw–Hill, New York.

Ralston, A. (1965b). Rational Chebyshev approximation by Remes algorithms, *Num. Math.*, **7**, 322–330.

Ralston, A. (1967). Rational Chebyshev approximation. In A. Ralston and M. S. Wilf (Eds), *Mathematical Methods for Digital Computers*, Vol. II, Wiley, New York.

Ralston, A. (1973). Some aspects of degeneracy in rational approximations, *JIMA*, **11**, 157–170.

Reddy, A. R. (1978). Recent advances in Chebyshev rational approximation on finite and infinite intervals, *J. Approx. Th.*, **22**, 59–84.

Reid, J. K. (1967). A note on the least squares solution of a band system of linear equations by Householder reductions, *Comp. J.*, **10**, 188–189.

Remes, E. YA. (1934a). Sur un procédé convergent d'approximations successives pour déterminer les polynômes d'approximation, *Comptes Rendues,* **198**, 2063–2065.

Remes, E. YA (1934b). Sur le calcul effectif des polynomes d'approximation de Tchebichef, *Comptes Rendues,* **199**, 337–340.

Rice, J. R. (1960). The characterization of best nonlinear Tchebycheff approximation, *Trans. A.M.S.,* **96**, 322–340.

Rice, J. R. (1962a). Tchebycheff approximation in a compact metric space, *Bull. A.M.S.,* **68**, 405–410.

Rice, J. R. (1962b). Chebyshev approximation by exponentials, *J. SIAM,* **10**, 149–161.

Rice, J. R. (1964a). *The Approximation of Functions,* Vol. I, Addison-Wesley, Reading, Mass.

Rice, J. R. (1964b). On the existence and characterization of best nonlinear Tchebycheff approximations, *Trans. A.M.S.,* **110**, 88–97.

Rice, J. R. (1967). Characterization of Chebyshev approximation by splines, *SIAM J. Num. Anal.,* **4**, 557–565.

Rice, J. R. (1969). *The Approximation of Functions,* Vol. II, Addison-Wesley, Reading, Mass.

Rivlin, T. J. (1969). *An Introduction to the Approximation of Functions,* Blaisdell, Waltham, Mass.

Rivlin, T. J. (1974). *The Chebyshev Polynomials,* Wiley, New York.

Rivlin, T. J., and Shapiro, H. S. (1960). Some uniqueness problems in approximation theory, *Comm. Pure and Appl. Math.,* **13**, 35–47.

Rivlin, T. J., and Wilson, M. W. (1969). An optimal property of Chebyshev expansions, *J. Approx. Th.,* **2**, 312–317.

Robers, P. D., and Ben -Israel, A. (1969). An interval programming algorithm for discrete linear L_1 approximation problems, *J. Approx. Th.,* **2**, 323–331.

Rogosinski, W. W. (1955). Some elementary inequalities for polynomials, *Math. Gaz.,* **39**, 7–12.

Runge, C. (1901). Uber empirische Funktionen und die Interpolation zwischen aquidistanten Ordinaten, *Zeit. fur Math. und. Phys.,* **46**, 224–243.

Schoenberg, I. J. (1946). Contributions to the problem of approximation of equidistant data by analytic functions, *Quart. J. Appl. Math.,* **4**, 49–99.

Schoenberg, I. J. (1965). On monosplines of least deviation and best quadrature formulae, *SIAM J. Num. Anal.,* **2**, 144–170.

Schoenberg, I. J., and Whitney, A. (1953). On Polya frequency functions III, *Trans. A.M.S.,* **74**, 246–259.

Schultz, M. H. (1969). L^∞ multivariate approximation theory, *SIAM J. Num. Anal.,* **6**, 161–183.

Schultz, M. H. (1972). Discrete Tchebycheff approximation for multivariate splines, *J. of Comp. and Sys. Sci.,* **6**, 298–304.

Schultz, M. H. (1973). *Spline Analysis,* Prentice Hall, Englewood Cliffs, New Jersey.

Schumaker, L. (1968). Uniform approximation by Chebyshev spline functions II: Free knots, *SIAM J. Num. Anal.,* **4**, 647–651.

Schumaker, L. (1969). Approximation by splines. In T. E. Greville (Ed.), *Theory and Applications of Spline Functions,* Academic Press, New York.

Scott, P. D., and Thorp, J. S. (1972). A descent algorithm for linear continuous Chebyshev approximation, *J. Approx. Th.,* **6**, 231–241.

Singer, I. (1970). *Best Approximation in Normed Linear Spaces by Elements of Linear Subspaces,* Springer, New York.

Spyropoulos, K., Kiountouzis, E., and Young, A. (1973). Discrete approximation in the L_1 norm, *Comp. J.,* **16**, 180–186.

Stiefel, E. L. (1959). Uber diskrete und lineaire Tschebyscheff-Approximationen, *Num. Math.,* **1**, 1–28.

Stiefel, E. L. (1960). Note on Jordan elimination, linear programming and Tchebycheff approximation, *Num. Math.,* **2**, 1–17.

Stoer, J. (1964). A direct method for Chebyshev approximation by rational functions, *JACM*, **11**, 59–69.

Szego, G. (1939). *Orthogonal Polynomials*, AMS Colloquium Publications, No. 23.

Taylor, G. D. (1973). Uniform approximation with side conditions. In G. G. Lorentz (Ed.), *Approximation Theory*, Academic Press, New York.

Tornheim, L. (1950). On n-parameter families of functions and associated convex functions, *Trans. A.M.S.*, **69**, 457–467.

Usow, K. H. (1967a). On L_1 approximation I: computation for continuous functions and continuous dependence, *SIAM J. Num. Anal.*, **4**, 70–88,

Usow, K. H. (1967b). On L_1 approximation II: computation for discrete functions and discretization effects, *SIAM J. Num. Anal.*, **4**, 233–244.

de la Vallée Poussin, C. J. (1911). Sur la methode de l'approximation minimum, *Soc. Sient. de Bruxelles, Ann. 2me partie, Memoires*, **35**, 1–16.

de la Vallée Poussin, C. J. (1919). *Leçons sur l'Approximation des Fonctions d'une Variable Réelle*, Gauthier-Villars, Paris.

Veidinger, L. (1960). On the numerical determination of the best approximations in the Chebyshev sense, *Num. Math.*, **2**, 99–105.

Wagner, H. M. (1959). Linear programming techniques for regression analysis, *J. Amer. Stat. Assoc.*, **54**, 206–212.

Walsh, J. L. (1931). The existence of rational functions of best approximation, *Trans. A.M.S.*, **33**, 668–689.

Watson, G. A. (1975). A multiple exchange algorithm for multivariate Chebyshev approximation, *SIAM J. Num. Anal.*, **12**, 46–52.

Watson, G. A. (1979). The minimax solution of an overdetermined system of non-linear equations, *JIMA*, **23**, 167–180.

Weinstein, S. E. (1968). *On the characterization and uniqueness of best uniform linear approximation*, University of Utah Report.

Weinstein, S. E. (1969). Approximation of functions of several variables: Product Chebyshev approximations I, *J. Approx. Th.*, **2**, 433–447.

Weinstein, S. E. (1971). Product approximations of functions of several variables, *SIAM J. Num. Anal.*, **8**, 178–189.

Werner, H. (1962). Die konstruktive Ermittlung der Tschebyscheff-Approximation in Bereich der rationalen Funktionen, *Arch. Rat. Mech. Anal.*, **11**, 368–384.

Werner, H. (1963). Rationale-Tschebyscheff-Approximation, Eigenwerttheorie und Differenzenrechnung, *Arch. Rat. Mech. Anal.*, **13**, 330–347.

Werner, H. (1967). Die Bedeutung der Normalitat bei rationaler T-Approximation, *Computing*, **2**, 34–52.

Werner, H. (1970). Tschebysheff-Approximation with sums of exponentials. In A. Talbot (Ed.), *Approximation Theory*, Academic Press, London.

Werner, H., Stoer, J., and Bommas, W. (1967). Rational Chebyshev approximation, *Num. Math.*, **10**, 289–306.

Wolfe, J. M. (1974). L_p rational approximation, *J. Approx. Th.*, **12**, 1–5.,

Wolfe, J. M. (1975). Discrete rational L_p approximation, *Maths. of Comp.*, **29**, 540–548.

Wolfe, J. M. (1977). Existence and convergence of discrete nonlinear best L_2 approximations, *J. Approx. Th.*, **20**, 1–9.

Wolfe, J. M. (1979). *On the convergence of an algorithm for discrete L_p approximation, Num. Math.*, **32**, 439–459.

Wulbert, D. (1971). Uniqueness and differential characterization of approximations from manifolds of functions, *Amer. J. Math.*, **93**, 350–366.

Young, J. W. (1907). General theory of approximation by functions involving a given number of arbitrary parameters, *Trans. A.M.S.*, **8**, 331–344.

Index